UE1:

UE1:

Departure

T. M. Ely

ISBN: 979-8-9855249-4-9 Paperback
ISBN: 979-8-9855249-5-6 Hardcover
ISBN: 979-8-9855249-6-3 eBook

Library of Congress Control Number: 2022907590

Cover design by: GetCovers

Printed by KDP, in the United States of America.

First printing edition 2022.

TDF Books
P.O. Box 100365
San Antonio, Texas, 78201

For my mother, for all the good times and the bad, for there is an everlasting bond of love between us.

A thought to ponder....

When people ask how your day is, my response is "better than some worse than others." Covers just about anything.

T. M. Ely

Hidden Easter eggs....

As with the first book, there are several hidden messages within the book, enjoy finding them, as I did when placing them.

Prologue

February 4th, 2008.

Rogers was removed from OTS so that he could help the UE1 team fix the computer issues along with some instrumentation issues with the XY33. After all, he built the prototype of the computer which is to run all the systems on the ship.

Starkes was informed of the mishap with the transport that was to bring Rogers to Groom Lake. It was unclear on his condition; the only word was there was one member in critical condition with several individuals hurt.

With only experimental aircraft on the base tonight, it was not possible to fly out of Groom Lake tonight, he would have to drive to Nellis and with time being of the essence, Starkes took off out of the office.

With a quick stop at his living quarters to pick up his go bag, it was not long before Starkes was well on his way to Nellis.

Pulling up to the main gate, stopped by the MP at the entry point. Looking at the decal on the window and seeing it was an officer asked to see his identification, as it was required to verify all personnel. Starkes held up his ID. All checked out, the MP gave a salute to Starkes and waved him through.

Proceeding to the base transit center Starkes was soon getting checked in. Waiting to board a military flight, one which would get him to Offutt AFB in Omaha Nebraska. One step closer to meeting with Rogers at the hospital in Columbus Nebraska. Starkes was busy playing out the events of the day in his head.

About twenty minutes passed before Starkes was called to get on the transport, a C-17 Globemaster. It would be just a bit over two hours of flight time to get to Omaha, he was looking at getting on the ground at about 2300 hours local time and then getting a room with Base Lodging, it was looking closer to midnight before he would be able to go to bed.

Having set the alarm for 0700 the morning was upon Starkes before he knew what happened. Calling down to the front desk, getting some information for a car rental. The drive would be about an hour and a half to get to Columbus, where Rogers was located.

After a short ride to the rental company and a quick bite to eat the time was fast approaching 0900 before Starkes was able to get on the road.

Chapter 1

February 5th, 2008, Columbus Nebraska, 1116 hours.

"Well, son I'm glad to see that you are still with us," Starkes said to Rogers.

Rogers was now becoming more aware of where he was, just not too sure why.

"Where exactly am I?" Rogers asked.

"Columbus Community Hospital, Columbus Nebraska. You were in an aircraft accident."

More of the events from the past few days were starting to become clearer for Rogers.

“Well, I hope that you will be able to clear something up for me?”

“Fire away.”

“Why was I taken out of OTS?”

“That’s an easy one, we have a small problem.

One that I’m hoping, well betting that you will be able to help us with. Besides, you already completed basic training. The doctors said that your CAT scan was normal, and you should be able to be discharged in a few days.”

“What’s the problem?”

Starkes looked around the room, with the door closed and only the two of them in the room he felt it would be ok to divulge some of the details.

“We sent an aircraft on a test flight, and it exceeded the ability of some of the instruments. We were hoping that you would be able to solve the problem in a short amount of time.”

“I don’t understand.”

“Rogers, we brought you onto our team for your computer and engineering skills, son, we are building a transport vessel that will be able to take us far away from this planet and deliver us to some other areas of the known, or unknown universe. We have set a launch date for July 4th of this year. Our first experimental fighter was on a test flight a few days ago and we had some unique challenges come up. Ones that I’m depending on your abilities to solve. Your computer is going to be the operations mainframe on the ship, which we already have built, but there seem to be some bugs in our prototype."

"You reproduced the computer I built at MIT?"

"Yes, using the specifications of the patent you had filed."

"Is there nothing protected in this country?"

"When it comes to the UE1 program, no, there is nothing that we are not exempt from using."

"Let me guess you are not able to get the computer to recycle the quantum physics contagion and adapt it to reality."

"Hell, son to be honest with you, I don't have a damn clue as to what you just said, but that sounds about right to me."

"I'm sure it is, there are a few parts I left out of the specifications that I summited to the patient office just for this type of reason."

"Well, I think that answers one question I have."

"Yeah, what's that?"

"If your brain was still in one piece."

"I do believe it is."

"Any more questions? Or are we good in till we get you back to the project site?"

"Yes, I have a few questions. Will I be able to see Michelle?"

"I'll do you one better than that. We also recruited her for this project, she will be joining the team in a few months."

"How? She's not in the military."

"We have a lot of civilians in the program, she thinks she is going to work at the FBI in the cyber crimes division. Anything else?"

"Yes, one more question. There was a guy on the aircraft that was helping me out, Woods was his name, how is he?"

"One of the nurses had said that you were asking about him. He was admitted in critical condition, but they said that he was going to be able to make a full recovery."

"So, tell me, why are you so interested in this guy?"

"As I said earlier, he was extremely helpful to me on the flight. Everyone else just seemed to give me attitude."

"Well, I just might have to look into that."

"Whatever do you mean?"

"Nothing, get some rest. I'm going to see about making your travel arrangements, so we can get you to your destination. Are you going to be alright with flying?"

"Yes, I think so."

"That's good to hear."

Starkes excused himself and left the room. Once more all to himself, Rogers started to ponder what was just told to him. Well knowing that he will be with the love of his life, Michelle. How much longer must he wait to see her once more?

Looking over to the side of the bed, on the table there was a radio, turning it on. *Please Don't Go, by KC and The Sunshine Band.* Was playing on the station that it was set to. It was an old song, one that he was not aware of but it had such a deep meaning and a whole new set of emotions ran all through him. Bringing a tear, and longing to be with her, he could only hope that she knew what she means to him. Rogers was moved deeply by the song and once more knew he had made the right decision to marry her.

Drifting off to sleep Michael Rogers was ready to start anew, with Michelle joining the program he felt so much relief and peace. With the love of his life entering his last thoughts before he was out, he would dream as no one has before.

February 8th, 2008, 0910 hours.

Walking around his room crawling the walls one might say, Rogers was ready to leave.

The charge nurse, Julie entered his room.

"Well Michael, it seems that the Doctor has cleared you to be discharged, and able to fly."

"Great, I didn't know how much longer I would be able to stay cooped up in here."

"Well, they are releasing you at about noon today. I thought you might want to take a short walk."

"A walk to where?"

"Oh, just down the hall, come on, it will do you some good."

Following Julie out of the room and down the hall, they stopped outside of another patient's room.

"Whose room is this?"

"Go on in you'll see," Julie said quite mysteriously.

Rogers slowly entered the room. There on the bed was Woods.

"Hey, I've been asking about you."

Woods responded slowly. "Well, I guess I've been right here."

"I was worried about you; I was trying to find you when we crashed. All they pretty much told me was that you were pretty banged up."

"That's what I hear, I don't even remember what exactly happened, the last thing I remember was telling the crew about the situation then I was flying through the cabin like a rag doll."

"Well, I'm glad to see that you are going to be ok. You were the only one that was nice to me, and I just wanted to say thanks. They are releasing me in a few hours, it seems that my travel arrangements are going to require some more flying." Rogers said with a small laugh.

"Hopefully, the landing will be a little smoother than the last one."

Now both were laughing.

"Well Rogers take care of yourself, I'm not always going to be there to feed you, you know."

"Just so you know that five-star meal you gave me didn't stay down for long."

"Well, maybe our paths will cross sometime again," Woods replied.

With Rogers saying his goodbye.

Thought to himself as he walked back to his room. *"If only he knew, if only."*

The time moved incredibly fast, it was already 11:45, and Julie had been by with his discharge instructions twenty minutes ago. Just waiting for the customary wheelchair ride out.

Entering the room there was an orderly with a wheelchair.

"Are you ready sir?" He asked.

"That would be an understatement."

Out the front door, Rogers went to a waiting cab. The driver greeted him.

"I have instructions to take you to the airport. Would that be correct sir?"

From the travel instructions that were delivered to his room by a military messenger yesterday, he was to proceed to the airport where he was to check in with the information desk and to have Captain Maxwell paged.

"Yes, that's correct. How long is the ride?"

"It's a short one, about ten minutes or so. Why are you in a rush?"

"No, I just need to be there about twelve-thirty, and I'm starving."

"Oh, well you got plenty of time. Would you like to stop off and get a bite?"

"That would be fabulous."

"What are you in the mood for?"

"Are there any sandwich shops nearby?"

"There's a Subway by the airport. How's that?"

"That will do."

Within a few minutes they were inside, each was getting a sub to go.

Arriving at the airport with fifteen minutes to spare. Rogers paid the fare and stopped to eat before entering the terminal building.

Once inside Rogers located the information desk.

"Can I help you, sir?"

"Yes, I need to have someone paged please."

"Yes, sir, who are you needing to have paged?"

"Captain Maxwell."

"Oh yes he had stopped by, said you would be looking for him."

Throughout the airport's concourse, you heard the page.

"Captain Maxwell your party is ready at the information booth."

After a brief time, Captain Maxwell appeared.

"Captain Maxwell, I presume?"

"Yes, and I take it your Rogers?"

"Yes, Sir I am."

"Great. Do you need to make a restroom stop before we board the plane? It's going to be a few hours before we get there."

"I'm good right now, I don't mind the lavatories on the planes if I need to go."

"If you say so, but I've not met anyone yet who is that comfortable with the bag."

“Barf bag?”

“No, the in-flight relief bag.”

Now more confused than ever, Rogers just brushed it off, not thinking about the commit.

As the two officers were walking down the jetway, then out a door onto the tarmac.

“Where are we going?” Rogers asked.

Maxwell pointing over to an F-15 fighter.

“What the hell?”

“That’s our ride.”

Over by the aircraft, there were two enlisted maintenance workers. They were sent down to help with the recovery of the C-130 that Rogers was on a few days ago.

“Good afternoon, Captain, Lieutenant, we are all set here.

“Good, would one of you get the lieutenant into the G-suite and helmet that’s in the travel pod. Then cover the basics and get him strapped in, he has never been in a fighter before.”

“No problem, sir, lieutenant if you would come with me, sir.”

Rogers followed him over to the side of the aircraft. Airmen Casey retrieved the G-suite and helped Rogers put it on.

“Just climb up the ladder and get in the seat, I’ll be up, right after you to show you the controls and get you strapped in.”

“Ok.”

Rogers was sitting in the seat, Casey brought up the helmet. Showing Rogers, the basics and securing him in. It only took a few minutes. Maxwell was now in the front seat, completing his preflight steps.

"Can you hear me back there?"

"Yes, sir."

"Good, this will only be a few minutes and we will be on our way."

Maxwell having completed his preflight steps had contacted the tower, got the needed clearance, and communicated with the ground crew for engine start.

The roar of the engine and the vibration going through the aircraft were a little unsettling to Rogers.

"Is this normal?"

"It's very normal, I understand your position, but rest assured lieutenant all is just fine."

Maxwell had completed all the flight control checks and was ready to taxi to the runway. Having gotten the clearance and the instructions, Maxwell started moving the fighter jet down the taxiway.

"Well, this is a lot different than the last one I was on."

"Just wait till we take off; this one is going to throw you into your seat, and you will probably lose your cookies as well."

"What cookies, I don't have any."

"Your lunch, that's what the bag is for, if you have to throw up do it in the bag, not all over yourself or the aircraft."

"Fast mover 229, Tower, you are clear for immediate departure."

"Tower, Fast mover 229 copy."

"Is that the sign that we can go now?"

"Yes, it is, just have your bag ready."

The engines were producing an enormous amount of noise, the pitch sounded like a tornado, then the breaks were released, and the sudden excursion Rogers felt was incredible. Rolling down the runway with a bird's eye view was amazing, the rate of speed the aircraft was traveling at was unbelievable. Within a matter of seconds, they were airborne.

"How are you holding up back there?"

"Great, I just feel amazing."

"Good to hear, hang on tight, I'm going to get us up to altitude in a second"

"Omaha Center, Fast mover 229, clearance to 40, over."

After a short pause, came the reply.

"229, Omaha, you're clear for 40, over."

"Omaha, copy."

"Time to move," Maxwell informed Rogers.

Without any further delay, the aircraft began to accelerate and ascend to the higher requested altitude.

Now cruising at 40,000 feet and going Mach 1.2 about 900 mph, it would only take a couple of hours to reach their destination, Groom Lake, Nevada, or as most of the world knows it to be called Area 51.

"So, lieutenant, what makes you so special?"

"What makes you ask that sir?"

"Well let's just say it's not customary practice for the Air Force to fly around the country in a 30-million-dollar fighter jet picking up lieutenants."

"Point well-made sir, but I'm under very strict orders not to speak to anyone about where or what I'm doing."

"Well let me guess, this came from Colonel Starkes."

"No, it was Captain Starkes."

"No, I'm pretty sure it's Colonel, he seems to use whatever title suits the situation as he sees fit."

"Figures, ever since I met him a few weeks ago everything seems to be in the shadows with him."

"By the way, between you and me, I'm the only one of us that knows where we are going."

"Aren't we going to Nellis Air Force Base?"

"Close but no cigar."

"Well, that's where I was heading before the crash."

"Lt, we are going to Groom Lake."

"Where?"

"Oh yeah, you're from that younger generation, Area 51."

"What?" Rogers replied, somewhat surprised.

"You know, I'm sure you've seen all of those Hollywood movies about the secret military base in the desert and the aliens and stuff."

"Yeah, I've seen a few I suppose."

"Well for the great works of entertainment they are, they are very audacious and true."

Taking a few minutes to ponder what was said and let it sink in Rogers was starting to process the events and get the full picturesque. Adding it up, one plus one was not adding up to be two but more like four.

"Ok so I think I'm starting to get it; we are going to go off into the universe?"

"Yes, lieutenant we are. Now, why did I fly halfway across the country to retrieve you?"

"Well, it seems that whatever you guys are working on there seems to be a problem with it, and the captain or colonel seems to think that I will be able to fix the problem. Especially since it is based on my prototype design."

"What was it that you designed? Damn, you're only what about twenty-two years old?"

"Twenty-one to be exact. I built a computer."

"A computer? There's more to it than that, computers are a dime a dozen these days."

"Well, this one can take the quantum physics contagion and adapt it to reality, process the information to form a basis of the social perspective, and make a decision without input from the controller based on the nucleus of the degree of the quotation from the applied astronomical principals of applied space."

"Ok, I'm a pretty smart guy, but what the hell did you just say?"

"It's a super-smart computer."

"How big is this thing? I've seen the ship, and nothing sticks out looking like a big computer."

"It's bigger than I originally planned on. My final design ended up about 1.5 meters squared."

"That does not seem very large for what you just said."

"By your dime a dozen computers, it is."

"Touché." Said Maxwell.

"Now it's your turn captain."

"Just call me Cosmo, that's my call sign. My turn for what?"

"Like you said the Air Force doesn't fly around picking up lieutenants. Why you?"

"Sounds like a fair question. One I'm not sure I have the answer to. But I crossed paths with the colonel and was recruited to be the test pilot on the fighter aircraft and the chief pilot for the UE1 vessel."

"So, what does this UE1 look like?"

"It's the biggest damn thing you have ever seen outside of the building that it is in, the best way to describe it, it's somewhat like an aircraft carrier. But it's a flying ship."

"Have you taken it up yet?"

"No, No, it's not ready, but we did fly the XY33 the other night."

"What's the XY33?"

"The most advanced aircraft, or maybe I should say spacecraft ever made."

"Why do you say spacecraft?"

"Well, the test flight only called for a ceiling of 75 to 80 thousand feet and within a matter of seconds I was at 75 miles."

Pondering the answer over, Rogers concluded.

"That would put you into space, a low-level orbit."

"Yes, now you can see where some of the problems come into play, the onboard instruments were not ready, better yet they were not capable of handling the extreme speed and the agility of the craft, a few of them even stopped working."

"Ok, it's all making more sense to me now."

"What's that?"

"Why I was removed from OTS before I completed the course. The colonel had told me when I was in the hospital about some issues that I needed to address. I was thinking they were just about my computer, but now I realize he was also referring to the XY33."

"Well now this talk has been good for both of us, it seems that we both have gotten our questions answered," Maxwell said.

"It would appear so." Replied Rogers.

"So, it's time for you to sit back and enjoy the ride, it's not going to take very long to arrive at our final destination." Informed Cosmo.

Chapter 2

February 23rd, 2008, Groom Lake (Area 51), 1352 hours

Rogers was getting acclimated to the team, going through a streamlined orientation. Being amazed by most of the technologies that were being displayed. It was not very long before Rogers was cut loose. With a huge laundry list of issues that the colonel assigned to him to tackle; Rogers was set to meet with Chavez early in the morning, to see what he could do about fixing the bugs on the XY33.

February 24th

Rogers arrived early for his meeting with Chavez.

"Good morning, Lt.," said Chavez.

"It's going to be a fabulous Moring." Replied Rogers.

"Good to hear your enthusiasm, but what makes you say that?"

"I think I have a solution for the instruments worked out."

"Already, you haven't even seen the aircraft?" Chavez said quite surprised.

"I know, there was a complete diagram on the iPad thing they gave me to use. After reviewing the specs I'm pretty sure I know what the issue is"

Chavez assisted Rogers in adapting the instruments, along with a rewrite of the basic flight computer code. With the components reinstalled into the XY33, they were ready to inform the colonel.

Picking up the on-base communication set and dialing the colonel's office it was not long before they were passing the good news.

"That's incredible, let's get a test set up." Replied the colonel.

With the time already late in the day, there was not enough time to get another test flight set up for today, but tomorrow night, there would be another run of the night flight of the fireflies.

With the day ending Rogers set off to his living quarters, time to draft a letter to Michelle. The only love of his life.

"Michelle, I'm in the last few weeks of my assigned technical training school, more like my first year of high school. Every day that goes by it seems more like I'm the one giving out the instructions. One thing I must confess, it has opened my eyes to a much larger world that we live in. I'm looking forward to when we will be together once more. These last few months without you have been killing me from the inside out, but my newfound faith is keeping me strong. Yours forever Michael."

All mail had to go over to the base communications office for review. After it was approved for content, it would be forwarded to Nellis AFB for distribution into the USPS. Which made it hard to say anything personal, as you never knew who was going to read it.

With the time getting late Rogers was beaten and ready to go to bed. The day promised to be another early start along with a late night.

The alarm clock was buzzing in his head, almost like a bee right in his ear. Rogers dragged himself up and out of bed, looking at the clock. The clock showed the time of 0530 hours. "*Well, I best get in the shower.*" He thought. Chavez would be by at 0600 to pick him up.

Just like clockwork Chavez was in the driveway at 0555. *"Damn that man is something else."*

There it was the dreaded knock on the door.

"Hey, are you ready yet?" Chavez yelled out.

"Yeah, I just need a second."

Not much time had passed before Rogers was out the door and getting into Chavez's vehicle.

Colonel Starkes was pushing the team to its breaking point. There was so much that needed to be done, with only a few short months before the planned launch date.

Chavez and Rogers along with the other assistant ground crew members performed various power-on checks of the equipment in the XY33.

Wherever you looked, the hanger looked like the inside of a beehive, getting ready to swarm. People were coming and going at a fevered pace. The time was winding down, it would not be much longer before the second maiden voyage.

This time, there was a protocol in place to notify Colonel Hyatt over at Peterson AFB, the SecDef, and the President.

With the hanger empty of all non-essential personnel, Cosmo was seated in the XY33 going once more through the checklist.

"Tower control, this is X-Ray Yankee One, over."

"X-Ray Yankee One, authorization code, over?"

"Kilo, Kilo, Two, One, Zulu, over."

Noticing that the authorization code was not the same as the other night, Cosmo just chopped it up to over-cautious security.

"X-Ray Yankee One, you are cleared to proceed up to step ten, when you are at step ten advise, over."

"Roger tower."

"Tower, X-Ray Yankee One, ready to proceed to step eleven, over."

"X-Ray Yankee One, after hanger doors open continue with your checklist to step twenty-three, then advise, over."

Meanwhile up in the tower.

"Colonel Hyatt, this is Groom Lake, we have an operational test flight in progress, and the launch will be at 2110 actual sir."

"What are the intentions?"

"This will be a max speed, low orbit, estimated at 125 miles." Informed the tower controller.

"Ok, I will have the ball contained."

"Tower, X-Ray Yankee One, holding at twenty-three, over."

"Copy, X-Ray, you are cleared to proceed to the main runway, follow the green flashing arrow to the EOR marker, over."

"Copy tower, proceeding to the EOR, over."

Once more Cosmo was parked at the EOR. Soon all the checks would be complete. Soon very soon he would be on his way up into the early night sky.

"X-Ray Yankee One proceed to 15L hold at the marker. You have a departure time of 2110 actual. When the marker turns green make for an immediate takeoff, time is of the essence. You are cleared for max speed run and destination of 125 m, over."

"Tower, copy, max speed for 125, over."

What seemed like hours to Cosmo, was only a few short minutes to everyone else on base. There it was the red marker light changed to green.

Advancing the controls to full military power, in a blink of an eye, the XY33 was gone, completely out of sight.

With the rush of emotion and adrenaline going through his veins, Cosmo felt the excitement once more. With the pressure suit expanding, which helps force the blood to stay in his upper body the G-forces had minimal effect on his body. This allowed him to maintain consciousness. Cosmo was quickly approaching the 125 mark.

"X-Ray, how are your readings, over?"

"Tower, we are five by five, everything is working like a charm, over."

"X-Ray, that's good to hear, hold the position for instruct, over."

"Tower, copy, over."

"Sir, what do you want to do?" Asked the tower controller.

"Call Hyatt, advise him we are going to take a walk around the block." Stated Starkes.

"Yes, sir."

"Colonel Hyatt, this is Groom Lake. Shadow said to tell you that we are going to take a walk around the block sir."

"All the way around?"

"Affirmative sir."

"Well, I'm glad he has so much confidence in me, out."

The line went dead.

"Sir we are clear." Informed the tower controller.

“Good, pass on the good news to X-Ray, tell him to play tag with a bird and maintain speed to mask his identity to probing eyes.”

“X-Ray Yankee One, you are to take a walk around the park, tag along with a bird and maintain its speed, so you are not picked up by any other eyes, over.”

“Tower, copy we are to play tag along.”

It was not very long before a satellite was approaching near Cosmo, with the satellite moving about 17,000 miles per hour it was going to take Cosmo on a ninety-minute joy ride around the Earth.

“X-Ray Yankee One, we ran some numbers you should be out of our communications for about seventy-five minutes are so, happy trails space cadet, over.”

“Tower, copy, over.”

Meanwhile, Hyatt informed his controllers of a computer simulation that would be taking place.

“Just inform me of the actions and the steps you normally would take, do not push the panic button. Repeat do not push the button.”

The whole room acknowledged the order.

“Ding, ding cat two launch detected, section 37°14’06” N by 115°48’40” W, repeat cat two launch detected, section 37°14’06” N by 115°48’40”W.”

“Sir I have a launch detection; repeat I have a launch detection.”

“Understood.”

“I would be pressing the red panic button.”

"Good."

"Sir the bogie is moving at an extremely high rate of speed, the computer is barely keeping up with the track."

"That's ok, what's your reading?"

"One hundred- and twenty-five miles sir, and it seems to be at a standstill, sir."

Hyatt took the phone from the aid.

"This Is Colonel Hyatt."

"Colonel Hyatt, this is Groom Lake. Shadow said to tell you that we are going to take a walk around the park sir." Informed the tower operator at Groom Lake.

"All the way around?"

"Affirmative sir."

"Well, I'm glad he has so much confidence in me, out." The line went dead.

A brief time later.

"Sir the bogie is on the move, well I think it's the bogie, it seems to be on the same elliptical path as one of the Milstar satellites sir."

"Ok son, keep a track of it and advise if it changes its course."

"Yes, sir."

Colonel Hyatt returned to his desk. Picking up the phone and placed a call to Groom Lake.

"Groom tower control."

"This is NORAD, your dog is on a Milstar leash, and it should be on an invisible walk around the park."

"Understood sir, thanks for the update."

"Tower to Shadow, I have a message from NORAD."

"Go for Shadow." Came the reply.

"Sir the dog is on a Milstar leash looks to be on a walk, and he's blind, over."

"Great, call me when the dog is back in the kennel, over."

"Copy sir, over."

For the next hour and a half, Cosmo would be in the cosmos. He was getting one incredible look at the plant, one which very few people have ever seen. Keeping an eye on all the instruments, Cosmo's time was well occupied.

There was such a wonder of peace and tranquility all around him. Over to the right about the two o'clock position was the ISS, it appeared to be much larger than what he knew it to be. The stated specifications were listed at 239' by 356' but to Cosmo, it looked more like 500' by about 1000'. Trying to picture a football field against the station, it defiantly was going to take a lot of football fields to equal this.

"Well, whatever they did to the instrument cluster it seems to be doing its job." Reading relative earth speed, the navigation computer was able to accept a fixed or known point as a reference to figure the distance. There was one gauge that provided the crafts speed, another which informed of the outside temp, and one for the inside temp. *"Now I know*

why Starkes went through all of the trouble of getting Rogers out to Groom Lake."

The ISS was well out of view, but there were numerous other space objects to gaze upon.

"It's so easy to forget why one is up here, but I better not miss my stop."

According to the data being displayed on the screen, it was time to contact ground control.

"X-Ray Yankee One to Tower, do you copy, over?"

There was a slight static hiss sound over the headgear, but still no response. *"I'll give it a couple more minutes."*

"X-Ray Yankee One to Tower, do you copy, over?"

In the background, Cosmo could hear some music playing. *Oh, how fitting, it was David Bowie singing Space Oddity*

"Tower, copy. How's your walk going, over?"

"Tower, the walk has been expressly exquisite, but the dog needs to take a break. By the way, that's not quite the way I remember that song, over."

"Copy, X-Ray, we have the kennel open and ready for a slow return to the shelter, use the homing transponder to guide your return. We thought you might like the newer version, over."

There was a lot of laughter taking place in the tower.

"Copy, Tower, what is the requested speed?"

“X-Ray, keep it under Mach 3, we would like to view the return flight on the screen, your departure speed was a little over the top, over.”

“Tower, copy, over.”

It only took about ten seconds to reach the 125 mark, but it was going to take a lot longer to return.

Before too long Cosmo was on the final approach. Easing the craft down ever so gently, the rear wheels touching the surface first, then the front wheel. Slowing down to taxi speed, feeling like he was not even moving at all, but he was.

Making the last turn, Cosmo was guided into the hanger, with the doors closing as soon as he was tucked inside.

With the aircraft chocked and receiving the hand singles from Chavez, Cosmo went through the shutdown procedures.

Picking up the phone and pressing the coded button which would ring to the colonel’s quarters, the tower controller was completing the requested task.

“Starkes!”

“Sir this is the tower, the dog is back in the kennel.”

“Good, where there any stated issues with the walk?”

“No sir.”

“That’s good to hear, I’m on my way.”

Starkes arrived at the hanger a brief time later. Going through the standard security measures before he was able to go inside and meet with Cosmo.

Walking into the main hanger bay, Starkes was approaching a group of men. One which included, Cosmo, Chavez, Rogers, and a few others.

"Well Maxwell, how did it go?" Asked the colonel.

"It was like a dream, I don't know what all Rogers and Chavez cooked up, but the improved flight computer along with the new gauges all worked like a charm. I would say she's ready."

"That's great to hear. So, you two, what did you do to fix the issues?"

"Well, it mostly was Rogers, I just did the heavy work." Replied Chavez.

"Well?" Said Starkes, looking at Rogers.

"It was pretty easy the flight computer was designed for an aircraft that flies in the earth's atmosphere, so I just did a basic rewrite of the program to use air pressure as an indicator as to the location of the craft, and we adjusted a few gauges to compensate for the different resistance."

"So then do we have an agreement? Are we ready to complete the assembly of the remaining craft?"

They all looked at each other for a consensus. Cosmo was first to speak.

"The way she responded, and feels, I would say yes."

"Chavez, Rogers, any thoughts?"

Both men responded.

"No sir."

Alright then, Chavez I need for you and your crew to go out to Skunk Works, Plant 42 in Palmdale they have been manufacturing the main components for some time. Take the updated specs on the changes that were made. Let's get the assembly completed on the squadron's aircraft. They should be able to start shipping us two birds a week."

"Rogers I need for you to get the computer on the UE1 operational. We need to have a complete system check on all the functions of the ship so we can start getting the bugs worked out."

"Understood sir."

Chapter 3

February 26th, 2008, Palmdale Ca., 1652 hours

Chavez and his team of six were just getting to the assembly plant. Having left a little bit before lunch, the team had made a stop on the way to eat.

Their instructions were to meet with the contractor liaison as soon as they arrived.

The first order of business would be to find the production building for Lockheed.

Once locating the right area, the team parked and started walking toward the entrance.

Entering through the famed entrance plaza, where an F-117 Nighthawk is on display.

"Looks just like the ones we had at Groom Lake a few years ago." Stated Tony.

Once inside the reception area, they were greeted by Beth.

"Welcome, you must be the team from GL that we have been expecting?"

Chavez replied. "Yes, we are, and I believe we are to meet with Roy Hodges."

"Yes, he has been asking if you have arrived yet. Let me page for him, if you would like some refreshments or the restroom it is located just off to your right, and there is a sitting area there as well. I'll send him down to you when he gets here."

"Thank you." Said Chavez.

Before Chavez could even think of anything else to say the team was off, seems a restroom break was in order.

Beth called down to Roy's office. "Sir the team from GL is here, they are down in reception."

"Thank you, Beth, I'll be right there."

Not much time had passed before Roy made his way into the reception area. As he entered the room.

"Hello, I'm Roy, how are you guys doing?"

"We're rather good, a little tired from the drive but anxious to get started in the morning. I'm Steve, that's Tony, Jack, Rex, Louie, and Lynda." Said Chavez pointing to each person as he said their names.

"It's great to meet you all, we have all been tremendously excited around here to finally get to complete the assembly of the XY33. So how was the testing?"

"For the most part, it was uneventful, a few bugs here and there and a few modifications that were needed."

"That's good to hear, well we have been stockpiling the airframes and all of the major components for the past couple of years, once we start full production, we will be able to kick out no less than two per week and we are hoping to get four a week."

"Well, that will make our commander very happy." Informed Chavez.

"There are a few major components we are lacking, the engines, and the fuel cells. Do you have a delivery date for those items?" Asked Roy.

"They should be here by the end of the week." Informed Chavez.

"Excellent, well let me show you to the conference room where we will meet in the morning say 0900."

“That should be fine, gives us plenty of time to get ready and get some breakfast.”

“There’s no need to get breakfast, we will have a variety of food options in the cafeteria.”

Looking over the team you could see some pretty big smiles appear on their faces.

Following Roy, the team set off for the conference room. Once inside the room, Roy gave each team member a temporary I.D. badge.

“This will allow you access into the building and the administration area including this room. This is where we will meet in the morning. You will also have access to the cafeteria on the subfloor, you can access it through the elevator just outside on the right.”

“What about the production area?” Inquired Chavez.

“No, you will need to check in with the security detail each day and receive the assembly floor access cards. No one is given access without going through security. Each time someone enters and exits the facilities they must go through a full personal check. You know to make sure they are not bringing in any contraband or taking out any confidential items.”

“We understand, we have the same kind of procedures in place as well.”

After showing them the basic layout of the area Roy gave them directions to the lodging facilities.

Having departed and leaving to go get checked in for their stay. The team had decided they would meet up in thirty minutes to go for some dinner.

Everyone had assembled in the lobby of the hotel. The look on their faces said it all, they needed a good steak dinner. After asking the clerk for some local recommendations, they headed off to get a satisfying meal.

Top of the list was the Texas Cattle Company, just down the street in Lancaster.

The place had that home look and feel, a little rustic, but the staff was very friendly and kept the drinks filled, the food was great and well-priced for the serving size.

Now that all had their bellies full, and the day had drawn to an end the team was more than ready to get some rest.

Chavez felt that he had just gotten to sleep when the alarm went off. Picking up the phone, called each team member's room to make sure they were getting up.

With the team scheduled to meet in the lobby at 0815, this would allow them ample time to eat before meeting with Roy at 0900.

Chavez was first to arrive in the lobby, next to arrive was Lynda, with the rest of the team showing up a few minutes later.

The drive over to Lockheed would only take about five minutes.

Once back inside, they went unrestricted to the cafeteria. Looking around, it was a large area, most of the tables were occupied by Lockheed employees. There appeared to be a few contractors in the mix as well. The cafeteria room looks as if it has available seating for about five hundred employees, it seemed to be a large area.

"Well, this must be the best place in town to eat, damn look at all of the people in here." Louie pointed out.

"Well, at least it smells good anyway." Stated Rex.

"All right, all right let's get in the serving line, time's a-wasting." Chavez implied.

Going through the serving line, the selection of items to pick from was just incredible. There were three meat items, three choices of eggs, pancakes, french toast, and waffles. There were at least five different kinds of fruits and several choices of bread and beverages.

After gathering trays and locating an empty table in one of the back corners the team posted up, somewhat to themselves.

Basic small talk was taking place amongst them as they ate their meal.

With the time nearing 0900, it was time to depart and head up to the conference room for the team debrief.

Upon entering, Roy was there with a few other personnel.

"Good morning." Stated Roy.

"Good morning to you as well," Chaves said.

"As requested in the communication from Starkes. Off to my left is Dr. Sinclair, he is the computer software developer, and next to him is John Epstein, engineering design, and last is Sam Peterson who is the overall project lead."

Chavez introduced his team to the three unfamiliar faces.

The room was closed and made secure for them to discuss in private the changes that needed to be made for the aircraft to perform as required.

Dr. Sinclair had numinous questions on the software changes that would need to be updated into the flight computer. Chavez tried to answer them as best as he could, but there were still a lot of questions that were beyond the scope of his understanding. Even after working directly with Rogers, on it, for the past few weeks.

"Where is the technician that made the changes?" Asked Dr. Sinclair.

"He was not able to attend as he is working on another high priority item." Informed Chavez.

"Can we reach him on a secure communication line?"

"That should be able to be arranged," Chavez replied.

"I'll take Chavez with me back to my office; we can go over the changes and try to reach this computer technician to get my questions answered so that I can get the updated info into the system." Dr. Sinclair proposed.

"Yes of course." Said Roy.

The group was led over to the elevator, Roy inserted his card into the master control and the elevator began to descend to the 2nd subfloor.

Chavez and Dr. Sinclair went the other direction up two floors to his office.

Once they arrived on the production floor, the team was amazed to see so many XY33 crafts in various stages of completion. Towards one end of the room, they looked completed, as you looked in the other direction there were just fragments of the airframe entering the production room.

Let's head over to the west end, those are the most completed birds to date." Stated Roy.

As before the team followed Roy, onto the production floor and down to the far end.

"So how long does it take from start to finish, to get one completed? Asked Jack.

"When we are in full production and capacity, it would take about a month." Replied Roy.

"Wow, a month?" Stated Jack.

"Yes, that's right when we are in full production mode. Right now, we are only at fifty percent, but that will all change this week. As soon as your team covers the approved changes with the production team members."

Now they were at the last bird on the line, one which looked as ready as the one back at Groom Lake.

"Well go ahead and look her over, the team leaders will be here in a few minutes." Informed Roy.

Each one went over to their area of expertise, to get a grip on the status and completeness of the aircraft.

After about ten minutes all the team leaders were present. With Chavez's team having looked over the airframe everyone went over to a production room for discussions.

"Well, what questions do you have for us?" Roy asked.

Lynda was the first to speak. "What's the status of the instruments, I noticed there were a few missing."

"Joe, would you like to feel that question? Please." Stated Roy.

"After we got the updated listing, we started removing them from all of the craft on the assembly line to expedite the required changes."

"What about the wheel assemblies? They are not the right size." Informed Rex.

"No those are just for production purposes they are changed out when the bird rolls out of the assembly line." Informed Webster.

"When will the fuel cells be inserted? Asked Louie.

"As soon as the engines are delivered from your project team," Janet said.

"How are you able to do any power on checks then?" Asked Jack.

"We have a master power cable that we can connect to the fuel cell harness." Said Janet.

With the various questions being exchanged between the two groups, it was becoming obvious that everyone had profound respect for each other in what they know and the work that they do.

With no other questions, the two teams split up into work groups based on their working knowledge and systems area of responsibility.

"So, Steve, tell me about this guy who rewrote my computer program. I've worked on that program, for close to three years." Said Dr. Sinclair.

"He's a young man that they pulled out from the academy a few weeks ago."

"What do you ever mean, he's only been there a few weeks and was able to rewrite the whole program."

With a slight pause in his response. "Well, yes."

"That's impossible, just to read the lines of code would take weeks, much less changing the formula."

"Well, I don't know the specifics, but he changed it in a few days."

"A few days, what in the hell do you mean."

"You will see."

Dr. Sinclair just sat there puzzled, more like dumbfounded. If you have ever seen a person with the deer in the headlights look, then you know how Dr. Sinclair was reacting.

"Dr. are you all, right?"

"Yes, I'm fine, just shocked. It would seem I should have known of this person and yct I don't."

"Well, he sent with me an external drive with the new program on it. Said something that you could just plug into the master computer, and it would incorporate the needed updated files onto your server. Which then could be transferred to the flight computers."

“That’s easy to do, but I still would like to speak with this remarkable gentleman.”

“I can try to reach him, but if he is working on the project that I think he is on it will be impossible to reach him.”

“But we still must try.” Insisted the Dr.

“How do I get a secured encrypted line?”

“Let me.” Dr. Sinclair said as he picked up the phone.

“This is Dr. Sinclair; I need to access a code one line.”

“Stand by please.” The operator replied.

There was silence on the line, for about thirty seconds.

“Your line is connected, sir.”

“Thank you.”

Handing the phone to Chavez.

“Just dial the number you want.”

Chavez picked up the receiver and dialed a number that connected him to Nellis AFB.

“Nellis operations. How may I help you?”

“Connect me to an AUTOVON line to Pentagon 5586.”

It only took a few seconds for the phone to begin to ring. A short time went by before a recording came on.

“Office of SecDef, please leave your message after the tone.”

“Sir this is Sargent Chavez with Team Shadow, requesting a relay to Rogers, Dr. Sinclair at Palmdale would like to speak with him ASAP.”

Hanging up the phone, looking towards Dr. Sinclair.

"Well let's see what happens."

"How long do you expect it to take, you didn't even leave a callback number."

"We don't need one, the SecDef will know with our code name."

"I felt that we worked in a veil of secrecy, but you guys are taking the cake."

"I would like to think that you would understand, considering where you work."

"I do, I do but it just gets to be a hindrance sometimes."

Webster and Rex went off to the wheel assembly shop. Joe and Lynda went to the instrument production area, it was not located in the production building, but one nearby. Both Louie and Jack were working with Janet, on XY33-1003. Tony stayed with Roy going over the final assembly schedule.

"When can we expect a shipment of the engine components?" Asked Janet.

"The last I knew, they were in the final stages of completion; I believe that they will be here by the end of the week," Louie informed.

Dr. Sinclair and Chavez were heading to the computer mainframe when there was a page for the Dr. Over the office paging system.

Stopping at the nearest desk, Dr. Sinclair picked up a phone and dialed zero.

"Operator, how may I help you?"

"Yes, this is Dr. Sinclair, I was paged."

"Yes sir, I'm connecting you now."

After a short pause.

"Your party is on Dr. Sinclair."

"This is Dr. Sinclair, who am I speaking with?"

"Dr. this is Lieutenant Rogers I was informed that you asked to speak with me?"

"Why yes, are you the man who changed my flight computer code?"

"I suppose that would be so, sir. What can I do for you? I instructed Chavez on the procedure to update the flight computers for the XY project."

"That I understand, but how were you able to go through all of that code in such a brief period and rewrite it?"

"Well sir, I'm not sure I'm at liberty to explain it."

"That sounds like some political B.S."

"All the same sir. What else can I help you with today."

"I understand the process but when the program transcends across the translucent disc drive, what are the ramifications to the memory capacity?"

"It's negligible, there is plenty of space left."

"Well, is there anything else that's important to know?"

"Not that Chavez would not be able to answer."

"How many simultaneous calculations can the system process?"

"I ran it with a thousand transactions a second without any issues."

“A thousand?”

“Yes, do you think we should increase it?”

“No, I just never heard of any computer running that many processes at once.”

“I use that number all of the time when I need to test a new method or procedure, it gives me a great reference point.”

“Indeed, it would.”

“Do you have any more questions I could help you with Dr.? There're some pressing matters here that need my attention.”

“Not that is immediate, but I would like to sit down with you and pick your brain sometime.”

“That day may come sooner than you could imagine.”

With that, the conversion had ended. Dr. Sinclair and Chavez continued their short journey to the mainframe.

“Well, behind that vault door is the mainframe, where we will need to insert the hard drive.”

Dr. Sinclair proceeded to insert his key card into the scanner. “*Please present retina for access.*” Preceding to present his eye to the scanner as instructed Dr. Sinclair had his eye scanned. “*Entry authorized, please stand clear of the doorway.*”

Within a few seconds, the vault door had opened, and with the entryway clear both men went inside.

Walking over to one of the terminal stations, Dr. Sinclair logged on.

Handing the external drive to the Dr., Chavez just stood back to watch.

It only took a few seconds to connect the drive. All at once, the room went what seemed like haywire, machines started to hum in one area and buzz in another, lights flashing all over the place.

"Is that normal?" Chavez asked.

"Yeah, it's just running some security protocol checks."

"I'm not used to seeing so many computers in one place."

"It can be somewhat overbearing to those that are not used to the everyday operations of large mainframe systems."

The screen displayed a message. *"Do you wish to proceed?"* Dr. Sinclair clicked on the box labeled yes.

Code lines started appearing on the screen, moving down amazingly fast. Just looking at the screen you would not be able to read any as it was moving too fast.

"Installation complete, you will need to reboot the system for installed changes to take effect." Came an automated reply.

"Well, there is an automatic reboot that will take place at midnight, so we will just have to wait till in the morning to run a compatibility test on the updates and see if it feeds into the downlink substations." Informed the Dr.

With the work done on the mainframe, the two went back to Dr. Sinclair's office.

Rex and Webster were in the wheel assembly shop inspecting the finished wheels. Rex was satisfied by what he was seeing. There was a massive amount of completed wheels stockpiled in racks going up to the ceiling.

"How many sets are ready to be installed onto the XY?"

Webster went over to a computer station. Typing in a few commands and getting the required information.

"It shows that there are ninety-five sets completed and thorough inspection."

"What is the total count to be delivered?"

Back on the computer.

"One set for each of the built aircraft, which is ninety-six, and another twenty-four complete wheel assemblies and an additional five hundred spare tires."

After looking around in the wheel assembly shop and talking with a few workers, Rex was ready to go back to the main assembly line. Needing to take a closer inspection of the strut assembly onto which the wheels were attached.

Joe was with Lynda in the instrument production shop. Sitting at one of the computer terminals, reviewing the diagram on the pulled instruments.

"Here where point 224 meets the capacitor, this needs to be rerouted to the main terminal bus and a resistor needs to be added to the circuit." Stated Joe.

"But that doesn't make any sense, the magnetic pull from the earth would change the stated readings."

"It's not going to get a reading from the earth, trust me it works."

"I don't understand how, but if that's what you guys want, then so be it." Said Lynda.

Pulling out a thumb drive Joe handed it to Lynda.

"Here is a diagram with the new instrument that we are needing to be installed in place of the speed indicator."

Looking it over and noticing, that there was an error on the readout.

"It's a digital readout, but there is an issue, what is the top speed you need on the readout? This has a number listed as 99,999!"

"That is correct."

"That doesn't make any sense. Why would you what an unattainable number?"

"Who says that it's unattainable?"

"Everyone."

"Well, my name is not everyone. Besides, we don't know the top speed yet, but let's just say it may be up there."

"No way, what's the recorded speed you've got so far?"

"They are still trying to crunch the number but it's over 20,000 mph."

"What?"

"It only took seconds to get to 125 miles up and he maintained speed with a satellite on an orbit around the planet."

Lynda was just standing there in total disbelief.

"Any more questions?" Joe asked.

"No, but I'm just a little puzzled."

"I know how you feel, you should have been there, one second we were looking at the XY on the runway, and the next it was gone, and I mean gone."

As the day went on the production on the next XY to be completed was taking shape. The last few adjustments would be made by the end of the week, just in time for the first batch of engines to be delivered.

With the teams all meeting back in the conference room to go over the day's progress, you could just feel the excitement and energy emerging from the room.

"Well, I just received word that we would be getting two sets of engines delivered overnight, they should be here by the time we start in the morning," Roy informed the group.

"That's a day ahead of schedule." Replied Janet.

"It seems that GL is very determined to get the production on the XY completed on time. How are we looking for the insertion of the engines?" Asked Roy.

"We are ready." Stated Janet.

"Great, well that will be the priority of the day."

As everyone was leaving Roy pulled Chavez aside. "What's this I hear about the other craft at GL?"

"What do you mean?"

"Oh, something about it traveling over 20,000 mph."

"I suppose that's true; we don't have a confirmed speed yet."

"What do we need to know about these engines? What dangers do they bring to my team and this facility?"

"If you're asking if there is any chance of a nuclear or chemical destabilization to be worried about, then your answer is no. We have been working and testing it for several years. What we do have is something revolutionary, it is a complete work of art."

"Well, that's where we have our difference, having worked with this end of the craft for the past several years, we are wondering how it is fueled as there are no fuel tanks. Which leaves a nuclear cell?"

"It's more like an Atomizer core drive."

"What kind of a drive?"

"Atomizer."

"Never heard of one."

"I know, it's never before been used or developed, until now."

"Well, this should be something to see."

Now that Chavez joined the rest of his team outside, they left for the hotel.

"What did Roy have to say?" Asked Rex.

"Not much he is a little concerned about possible contamination with the fuel cells and the engines. It seems the cat is out of the bag concerning the aircraft's speed."

"That probably was me, when I was working with Lynda on the new speed instrumentation." Joe said."

"Well, it was going to happen as soon as they set their eyes on the engines anyway."

"I don't understand how they developed the craft without knowing." Louie piped in.

"They worked off of the plans that were submitted from our team back at Groom Lake." Informed Chavez.

"Have you noticed that no one says Groom Lake, all they ever say is GL?" Stated Joe.

"Yeah, I've got that too." Said Jack.

"Well, I'm sure it has something to do with their security protocol. Now let's discuss more important matters. Where are we going to eat?" Asked Chavez.

"Pizza." Rex was the first to speak up.

"I'm good for pizza." Jack agreed.

"Well, if we agree, then pizza it is." Said Chavez.

By the time the pizzas made it to the table the team was just about famished. Once all of them had their fill, they were ready to get back and get some sleep, for tomorrow was going to be another long day with four engines to install and a preflight checkout to perform.

Chapter 4

February 27th, 2008, Groom Lake, 0031 hours

"How much longer before that plane is ready for departure?" Starkes was inquiring.

"The last crate is being loaded as we speak sir."

"What about the security detail?"

"They are ready, and I've confirmed with Nellis on the fighter escort."

"Great, so we are good for a 0100 hundred takeoff."

"Yes, sir."

As the time ticked by Starkes was pacing about like a father waiting for his daughter to return from her first date.

The security detail was aboard the transport aircraft, the doors were closed, and all was set on this end.

"Nellis tower."

"Nellis, this is Shadow, we are ready for departure."

"Shadow, we have the fighters on ready alert, five minutes out."

"Copy, proceed with a launch at 0100 Zulu, transmit final orders to the flight lead."

"Copy, Nellis out."

With that, both controllers hung up the communication device.

"Tower, Heavy One One Five, you are clear for engine start, over."

"Heavy One One Five, Tower, copy."

The aircraft commander was on the intercom with the ground maintenance crew chief.

"Ready for engine start."

"All clear sir."

Without any further delay, the aircraft had all three engines purring like a kitten. Going through their checklist the crew aboard the KC-10 were ready to taxi.

"Sir it's 0055." Stated one of the controllers in the tower at Nellis.

"Ok, let's hit the klaxon, wake them up."

Three loud siren chirps sounded in the flight line shack, along with the one in the pilot's ready room. Within seconds the maintenance troops were out the door, and a van was pulling up to the ready room. Four pilots and four WSOs were jumping into the van. Racing off to the alert aircraft.

Once the van arrived, pulled up to the first aircraft.

"169." Called out the driver.

Two men jumped out, running over to meet the crew chief.

This aircraft had already had a preflight inspection performed. Standing by on alert status, ready to go at a moment's notice. The pilots shimmed up the ladder, which the ground crew promptly removed as soon as they were strapped in. Just as soon as the ladder was clear the canopy was closed.

"216, and 173." The driver now called out.

Two more crews jumped out the back and took off for the aircraft.

The van was now pulling up to the last one. "116."

The last crew got out.

All the crews were in their assigned aircraft, all had preceded to start engines and were ready to taxi. Captain Bella was the flight commander for this group.

"Hey Rick, we should be getting the all-clear sign any minute, then we can go finish that card game."

"Tower, 216 you are clear to move to EOR, over."

"Well damn, I guess that game is going to have to wait a bit," Bella said to Rick.

"216, Tower, copy."

Captain Bella gave the hand gesture to the waiting crew chief the sign to remove the wheel chocks. Chief repeated the sign to the two troops that were under the aircraft. As soon as they were clear he took his crossed arms and raised them above his head holding up two fingers on his right arm and shaking them in a small circle.

The pitch of the engines increased as the power was applied. He moved both of his arms back and forth, took his left arm, and pointed in the direction that the aircraft needed to go, still moving the right arm back and forth. As the aircraft started to pass, he turned and gave a salute, which was returned by Captain Bella. The crew chief reached out with his left hand and brushed the wing tip as his aircraft rolled by with two fingers.

Looking down the flight line you saw the same process is repeated, 169 pulled out of its parking spot, then 173 and 116 took up the last position.

As they were moving down the taxiway.

"Get me a rundown on the weapons store," Bella asked the WSO.

Going through the systems.

"We have the LANTIRN online, packing four AIM-9s, along with four AIM-120s, and the 20mm has 500 rounds with every ten a tracer." Stated Rick.

Sitting at the EOR, all four F-15Es were waiting on their final instructions, expecting to be told to return and shut down.

"Tower, 216, go to secure mode on foxtrot, over."

“216, Tower, copy.”

Changing the settings on his radio unit Captain Bella called back into the tower.

“Tower, 216, copy the message in one part and expedite, over.”

“216, Tower, ready, over.”

“Your call sign is Shield, you are to meet up with Heavy 115, it’s a clean KC-10, call sign Packer, over the test range and escort to Palmdale, this is a top priority from SecDef. You are cleared for weapons hot upon meeting with Packer. You are to use whatever force necessary to ensure that Packer arrives unmolested to Palmdale, over.”

“216, Tower, copy.” Replied Bella.

“Lead to flight, change to Bravo over.”

After giving them a few seconds.

“Ok, flight make sure you keep your encryption on and stay on Bravo keep the traffic down, this is not an exercise, repeat this is not an exercise we are on a classified mission. We will be going weapons-hot in a short bit; we are clear to engage as needed. We are flying escort for a KC-10, over.”

“Tower, Shield, you are clear for a two-by-two immediate takeoff, over.”

“Shield, Tower, copy.”

“Ok, One Six Nine take my Three for a standard Two by Two, One Seven Three, and One One Six follow and expedite, over.”

216 took up position on the main runway with 169 pulling alongside at his Three o'clock position.

"169, ready." Stated the other pilot.

With both aircraft increasing their engines and holding on to the breaks.

"On my mark…. Three, Two, One, Mark."

Both aircraft took off in a violent thunder of force with the Pratt & Whitney F100-PW-229 engines producing 58,000 pounds of thrust it was only a matter of seconds before they were in the air.

Arriving at the designated rally point.

"Shield, Packer, do you copy?"

"Packer, Shield, you are loud and clear, what's your code, over."

"Tango, Tango, Whiskey, Four, over." Replied Bella.

"Good to have you here, we have a rush on our cargo, and we are to proceed without delay to target."

"Understood, Packer."

"We have a clear window at 40 and the side view is ten by ten."

Said the crew from Packer.

"Lead, One Six Nine take the Three, One Seven Three you got the top cover and One One Six keep the Six open, over." Bella directed to the other F-15 flight crews.

With each of the aircraft moving to their instructed position, the KC-10 was surrounded.

"Lead, team you are clear for weapons hot, call out any detected threat, the ROE (Rules of Engagement) is very clear, respond with whatever force is needed, over." Stated Bella.

Now Captain Bella was left to ponder. "*What in this world could they possibly have on there that it would call for this kind of protection over American soil?*"

Captain Bella, having been in the USAF for the past eight years has been on many deployments and training excursions, and never during that time has he ever had the authorization to go weapons hot, not even after 9/11, one still had to call in before you could activate your weapons.

"Whatever they are transporting must be of some importance," Bella said to the WSO.

"What makes you say that sir?"

"When have you known to fly with this much firepower and have a green light to take out any threat and not be in a war zone?"

"You make a good point sir."

"I just hope nothing enters our flight path, I damn sure don't want any screw-ups on this mission."

Starkes was still pacing back and forth in the tower.

"Get me a confirmation that Packer and Shield have met up?"

"Yes, sir."

Picking up a headset the controller punched in a code number that would connect him with Nellis Tower.

"Nellis, Sergeant Andrews speaking."

"Sergeant this is Shadow requesting a confirmation on the meet and greet."

"Stand by Shadow."

"Nellis Tower, Shield, over."

"Shield, Tower, over."

"Tower, how is the meet and greet going, over?"

"Shield, Tower, we have a shack and bake in the oven and on target for a 0230 dinner, over."

"Tower, Shield, copy, advise when the timer goes off, over."

"Shield, Tower, copy, over."

"Shadow, the dinner is in the oven, and it should be ready at 0230. Anything else I can help you with tonight?"

"No sergeant that will be all, thank you."

"Sir, Nellis reports that all is well, and they should be on the ground at 0230 Zulu."

"Good, buzz my quarters when they are on the ground."

"Yes, sir."

Starkes left to go to his living quarters. There was no way he would be able to sleep in till he knew that the transport was safely on the ground at the assembly plant in Palmdale.

"Sir I got a bogie 50 miles out moving right for us."

"Lead, Smoke investigate the bogie and get it out of our path."

"One Seven Three, copy."

Without delay 173 was in full afterburner, hitting Mach 2 in Three seconds, all you could see is the flame from the twin engines as it disappeared into the night sky.

Calling out on the standard civilian channels 173 was attempting to make radio contact with the unknown aircraft approaching their position.

"AF173 to unknown aircraft at FL290 on 180 heading, respond."

Repeating the message several times, finally, a response came through.

"This is AA6120 we receive you, over."

"AA6120, you need to change your course, request an immediate turn to 174, keeping FL290, over."

By this time 173 was completing a precision turn and pulling alongside the civilian aircraft just off the left side and matching the flight speed.

"Military aircraft 173, what's the nature of this requested change?"

"AA6120 time is of the essence, request the flight change ASAP."

At the same time, 173 turned the belly of his aircraft up towards the pilot, rocking it back and forth several times, showing that he meant business.

The captain of the civilian airline decided that it would be in his best interest to communicate with ATC and get a new heading.

"6120, LAX ATC, request a change in our heading to 174 at the direction of military flight, over."

"ATC 6120 you are clear at your discretion to 174, over."

AA6120 started to make a concise and steady turn to the new heading of 174. The F-15 pulled back and maintained a shadow on the airliner just to keep him in his targeting sights.

"Smoke, Lead, over."

"Go for lead, over."

"The bogie is on a new heading to One Seven Four, over."

"Copy, Smoke, hold for five then rejoin, over."

"Copy, Lead, over."

As AA6120 kept steady on the new course, 173 maintained for five more minutes to ensure that the instructions were entirely carried out. As the time had passed 173 turned to rejoin the flight group, increasing the aircraft's speed, just below the speed of sound.

Upon returning 173 resumed the top cover position, with the WSO checking for any unwanted aircraft in the near vicinity. Seeing all was clear on the scope as well as a visual inspection.

Without any further incidents, the flight group made its way to Palmdale.

After a brief taxi, all the planes were parked near the assembly building. There was a large security detail guarding the transport aircraft.

The crews from the fighter jets and the KC-10 were taken over to the VOQ, as soon as they had cleared the flight line area the overnight crew got busy unloading the engines and moving them inside the assembly plant.

Once the day shift and the team from Groom Lake arrived in a few hours, all would be in place for them to complete the next few XY33s.

Chapter 5

February 27th, 2008, Palmdale Ca., 0800 hours

With the team meeting up in the briefing room, the news was very quickly disseminated between them.

"Early in the morning, GL delivered a batch of engines for the XY. We should be able to have them installed today and ready for a test-fire of the complete system tonight. There will be a test pilot arriving here later today, so let's get the show on the road." Hodges informed the team.

With the newly arrived aircraft parts, there was plenty of work to be done today. Not much time had passed before the first main engine was installed, followed by one of the auxiliary control engines.

With the time creeping up on the lunch hour, the team decided to work on through the completion of the engine installation. Looking to complete the work by Sixteen Hundred hours.

The day proceeded without any major complications, which was very pleasing to Hodges and Chavez. With a self-imposed deadline nearing you could see that the work was nearly complete.

Now with the end of the workday upon them, the team had gathered around the next two XY's ready to roll from the assembly line.

"When are they going to proceed with the engine run test? Joe asked.

"I think that Cosmo will be here later today, so it will probably be later tonight. I should have an update soon from command." Chaves replied.

February 27th, 2008, Groom Lake, 1300 hours

Cosmo was in his living quarters packing a small travel bag, having been instructed to be at the Palmdale plant by nineteen hundred local time. He still needed to go see Elizabeth and tell her goodbye. He stopped by the engine shop to pick up the flight manuals and check in on the UE1. There was still a ton of work to get done before leaving.

Picking up the phone Maxwell called over to the Flight Operations office.

"Good afternoon, Sargent Spaatz speaking. How may I help you?"

"Sargent, this is Captain Maxwell I'm checking on the status of a fast mover to Palmdale. Has the order come down yet?"

"Affirmative sir, the bird is being prepped as we speak and should be ready for a 1400 departure."

"Great, I'll check back in shortly."

"Yes, sir." Came the reply from Spaatz.

It only took a few days to fix the issues with the computer on the UE1. Rogers was now calling it systems memory and relative theory, or SMART for short. With the scheduled departure for the UE1 set for this summer and the successful test flight of the XY33, it was now being placed into full production mode.

With a full squadron of aircraft needing to be assembled and spare parts bought onto the UE1 before the scheduled departure date, almost all the base operations were working around the clock, in support of the UE1 project.

Rogers was going to the end-of-day briefing, which was the midafternoon with the design and build team. The commander would also be present this time. Knowing that they all would be ecstatic with his update.

"Ok, what's the status on the propulsion system?" Starkes was asking.

Hansen spoke up. "Sir the modifications have been completed, and we are just waiting on the master computer system to be online so that we could proceed with a warm engine run."

"Captain Ely, where do we stand on the training program?"

"Sir all of the scheduled training has taken place and we are working to update the new XY33 manuals with the latest changes."

Looking over at Rogers it looked as if he had ants in his pants, barely able to keep himself seated.

"Ok, Rogers what do you have for us?" Starkes asked.

"Sir SMART is online and running a complete ship diagnostic and practical simulation test as we are speaking, I am expecting an update to my communicator any minute now."

"What the hell is smart and what communicator?"

"SMART is what I've named the computer. Systems Memory, And Relative Theory. Smart for short."

"And this communicator?"

"It is just something I put together so that I can access the system from wherever I might be on the ship."

Just as Rogers was completing his last statement there was a beep beep sound from his pocket.

"Well, Rogers please don't keep us in suspense any longer."

Pulling out what looked like a phone Rogers was looking at the data being displayed on the screen. Pressing one of the buttons, the device began to speak. "*What can I help you with?*" Speaking into the microphone.

"What is the status of the diagnostic test?"

"*Test completed at 1523 hours; no primary systems errors detected.*"

Upon hearing this news, Starkes asked. "Does this mean that we can proceed with a live-fire test of the systems?"

"From my point of view, I see no reason as to why we could not move forward with a self-power on check."

"Hansen?"

"As I said earlier sir, we are a go for an engine fire test."

"Where is Maxwell?" Asked Starkes.

Starkes just realized that Maxwell was not present at the meeting.

Elizabeth spoke up. "Sir he is on his way to Palmdale."

"Oh yeah, that's right, I sent him there for the XY33 test. They should have two birds ready to fly."

"Let's move forward and set full power on check for the 29th, 2100 hours. Only the section chiefs and their designated personnel should be on UE1. Let's identify the required personnel and dispatch the order by 1500 hours tomorrow." Starkes directed the team.

"If there's no other pressing information for me then I've got some other matters to attend to," Starkes said as he began to stand and then turn for the door, proceeding to leave the room.

Elizabeth was the first to speak. "Well Rogers, it is your computer that's going to be controlling this. So, what departments do you need?"

Pausing and pondering the question, Rogers was feeling totally on the spot.

"Well, I am not sure, I guess we defiantly need the Propulsion Section, Environmental, Navigation, and I guess someone from Security, for starters."

"Ok, Hansen, you should know whom you need from your team, I can put a few names together for Environmental, and Navigation. What about our Pilot?"

"Yes, that was a given I thought?" Rogers said.

As the list of names was taking shape, the list seemed to be getting quite lengthy. Although it is an enormous vessel, with thousands of personnel to be assigned to the program, considering that, the list did not seem so large at all.

"I'll process the dispatch order and inform the security team, as to the planned event," Elizabeth informed the group.

With the completion of the team lead roster, they were now prepared for a live systems check. The group disbanded for the day.

Elizabeth took off for the flight operations office, knowing that she should be able to catch Maxwell before he departed.

Upon arriving and approaching the door, Elizabeth spotted Maxwell.

"Are you trying to sneak off without saying goodbye?" She asked.

"Never in a million years," Maxwell replied.

The two of them reached out for a short embrace.

"Well, it's official, we are performing a full systems power check in two days. So, try not to get distracted on your little escapade."

"Why, are you going to miss me, Mrs. Ely?"

"Are you trying to patronize me, Mr. Maxwell?"

The two of them started laughing together.

With a few more minutes of small talk taking place, it was time for Maxwell to get situated. The time was fast approaching 1600 hours.

Elizabeth said goodbye. With a light peck from her tender lips and caressing the back of his head, Maxwell got became a little weak in the knees.

Upon seeing her walk back to her vehicle, Maxwell entered the flight ops. Proceeding to walk up to the duty sergeant.

"Good afternoon, sir, your bird is ready for an immediate departure. I just need you to approve the flight plan. So I can get it inputted into the system." Informed Spaatz.

Maxwell looked over the flight plan. Knowing that there would not be anything out of the ordinary, signed off and handed it back to the sergeant.

"Very well sir, the data should be on the tower control screen in a matter of minutes. By the time you complete your preflight, they will have the information."

"Great, I'll see you in a couple of day's sergeant."

"Very well, sir you can proceed to the aircraft, the airmen at the back exit will show you to your bird."

Maxwell headed to the rear of the building, carrying his travel bag with him. Inside there also was a new checklist for the XY's, that would be coming from the production line.

Once Maxwell performed the required walk-around of the aircraft and looked over the maintenance logbook. Seeing that all was in order, stored his bag in the travel pod.

The airmen secured the travel pod door and proceeded to help him get strapped into the ejection seat. Once the ladder was removed, Maxwell continued going through the preflight checklist.

On the radio set, Maxwell was given clearance for engine start from the control tower. Soon after, all the flight control checks had been completed. He proceeded to the designed runway for takeoff. It was not very long before Maxwell was airborne, on the way to Palmdale.

The flight was going uneventful, just the way any pilot likes it, no matter what they say back at the club. Not much longer and Maxwell would be on the final approach to Palmdale.

Going through the last few days and what would be taking place in the next few to come, Maxwell's time was kept occupied.

"Palmdale ATC, Faster mover 550, you are clear to turn to 115, at the outer marker, hold 150 feet, over."

"550, ATC, roger, over."

Maxwell turned the jet onto the final approach for the main runway. Not much longer and he would be parked near the assembly hanger for the XY33s.

Going through the shutdown procedures Maxwell proceeded to exit the aircraft. Upon his departure, Maxwell stepped into the waiting courtesy vehicle. Which would take him over to the main entrance, where he was to meet with Chavez and Hodges.

"How was your flight, Captain?"

"Uneventful, just the way I like it."

"Good, well I do believe we have a couple of birds ready for you."

"Fantastic, I did not want to make a trip for no reason, Chavez."

"Oh, don't be so melodramatic." Replied Chavez, as both men started to laugh.

"This is Hodges; he is the project lead here."

"It is a pleasure to meet you, sir."

"Same here sir, well we have you scheduled for a test run tonight at 2100 hours," Hodges stated.

"Sounds good to me, gives me plenty of time to get to my room, eat, and rest up a bit."

"We have a vehicle available for your use, there is a base map inside with the directions to the VAQ/VOQ. We will meet back here at 20:30." Hodges informed Maxwell.

With the pleasantries out of the way, Maxwell was looking forward to getting to his room to rest up, before the power-up tests on the twoXY33s that would be rolling off of the assembly line today. The first of many that would require his services before they could be delivered to Groom Lake.

Chapter 6

February 27th, 2008, Groom Lake, 1443 hours

"I don't give a rat's tail end, that is the personnel list that was summited to JCS, (Joint Chief of Staff) approved and the orders should have been processed by now. What is the damn holdup? I was expecting to have them here on Monday, now you are telling me nobody has been given orders, and one of the individuals has been released from active duty." Starkes, yelling into the phone.

"Yes sir, that is correct. Sir, all I know is what shows on my computer screen. The orders have not been processed; it still shows needing higher authority approval." Informed a clerk from the personnel processing center at Randolph AFB."

"Son, don't you move five feet from your desk, you will have a phone call from the highest authority in ten minutes." Starkes barked into the phone as he hung it up without waiting for a reply.

Yelling into the intercom to his secretary. "Get me the president on the line."

"Yes sir, at once." She replied.

Less than thirty seconds had passed when Alex buzzed the colonel.

"I have the president on the line for you sir."

Picking up the phone. "Starkes here sir."

"So, what do I owe the pleasure of this call to today, Colonel?"

"Sir, I'll cut straight to the chase. Someone seems to have dropped the ball on the personnel orders that were submitted to JCS. Randolph says that they still require higher approval, and one of the individuals has already been out-processed from the service. We were expecting them to be here on Monday."

"What the hell!"

"That pretty much was my response, sir."

"What do you need from me?"

"If you could get with the commander over at Randolph and inform him of the urgency on the personnel re-assignments. It would get the ball moving."

"Considerate it done."

"Thank you, sir, I knew you would feel the same way."

No sooner than Starkes had gotten off the phone with the president, the president's staff was getting connected to the commander at the Personnel Center at Randolph Air Force Base, located in San Antonio Texas.

"Please hold for the Commander in Chief."

"Mr. President, Colonel Steves is on line five, sir."

"Colonel, there seems to be a slight problem down there in your office."

"I am not following you sir; It is just a normal routine workday."

"Well, I've just been informed of some orders that have not been processed, there seems to be a hold-up on them still requiring a higher approval. These personnel orders are for Project Star Search, and it is having a detrimental effect on the outcome of this program. Also, one of the assigned individuals has been released from active duty. We need to see about a recall on them and get this done today. I don't care what it takes, get them in place by Monday of next. Any questions Colonel?"

"No sir, it will be done."

"Very good."

Then the line went dead.

Colonel Steves was looking at the computer monitor, typing in a few commands on the keyboard. Seeing that Star Search was assigned to Senior Airmen Devon. Upon getting this information it was just a matter of calling his extension.

Senior Airmen Devon was sitting at his desk, it had been seven minutes since that arrogant jerk had hung-up, on him. Thought that he is a fool to be sitting by waiting for his phone to ring, besides, he was already late for his break.

Just as Devon stood up to leave. Beep, Beep, Beep went his desk phone. *Thought that it was just a coincidence.* He reached down and picked it up anyway.

"Senior Airmen Devon speaking, how may I help you?"

"Airmen this is the Colonel report to my office ASAP."

And just like before the line went dead.

Gathering his composer Airmen Devon made haste to the Colonel's office. Entering the reception area for the Colonel, Devon met up with the Colonel's aide.

"Reporting as the Colonel requested."

"And who might you be?" Asked the aide.

"Senior Airmen Devon." Pointing over to some chairs.

"You may sit over there."

Picking up his phone and buzzed the Colonel.

"Yes, Jake."

"There is a Senior Airmen Devon here for you sir."

"Send him right in."

"Yes, sir."

Calling out to Devon, who didn't even have time to make it to the chair.

"You may go straight in."

Upon entering the office.

"Close the door behind you."

Devon closed the door as he completely entered the office, walking towards the desk to take up proper reporting procedures.

"Just skip the formalities. What the hell is taking place with the personnel orders for the Star Search project?"

"Sir I just received a call not long ago, I'm not even sure who it was asking about it. The computer still has it listed as needing a higher level of approval. So, there has not been any processing of the latest request."

"Just so you are clear and understand me. I just got off the phone with the president, so the authority is not going to get much higher unless God himself signs them. You are to contact each of them by the end of the day and report back to me that it has been completed. You are to inform them that they will have a hard copy sent to their command by COB tomorrow, they will need to be on station be COB Monday."

"Sir, there must be a hundred people on that list. What about the one who has been discharged?"

"You need to see what his status is and see if he falls under one of the recall actions. Why are you still standing here? Get your rear in gear."

Devon made like the wind and blew out of the Colonel's office faster than a bolt of lightning.

Walking back to his desk faster than when he went to the Colonel's office. *Damn, that went well, I guess he didn't need the whole ten minutes. How the hell am I going to get in contact with all those service members by today?*

Arriving back at his duty section Devon ran into his section chief, Sargent Reed.

"Devon your break should have been over ten minutes ago."

"Sargent Reed, I was."

Getting cut off from his explanation.

"I know, I know, I'm just pulling your chain a bit. Get a list of the names and commands, meet me in the Alfa conference room in ten."

Devon had the requested information in five and made his way to the conference room. Reed was already there along with some of his other team members.

"We are waiting for two more and then we will get briefed," Reed informed the group.

The time was approaching 1530, not leaving much time in the day as most commands released their personnel at 1645.

"Ok, guys, I'm going to make this short and sweet, we might have to stay a bit late today."

You could hear a few moans and groans in the room.

"Devon has a list of names, there are about a hundred, and we need to make personal contact with each of them today. Don't leave a message, when I say we are to speak with the individual I mean we are to have their verbal commitment. Is that clear?"

You could hear that the room got noticeably quiet knowing that Reed was serious.

"Yes, sir." Everyone replied about the same time.

"Each person's instructions are the same, they are to report to their new duty section at 1500 hours this Monday, yes, I know this is very short notice if they have families they will be sent for later. Their commands will get hard copies of their orders by COB tomorrow. If there is a military flight available, they will have priority, otherwise, they are to use civilian travel. Any questions?"

One hand went up.

"Yes, Jackson."

"What do we do if we are not able to reach someone?"

"Call to their current duty commander's office, inform them of a *Redball Action* to locate the individual in question and have them return the call, then move to the next name on the list. All right then, if your list is done check with your coworkers, no one is leaving until this is complete. Let's move people."

There was a fury of activity taking place, everyone was on their phones. One by one names were getting scratched off their assigned lists.

Sitting at his desk, Devon was making one of his assigned calls. Beep, beep, beep was the tone heard in his headset.

"Lieutenant Harper speaking; How may I help you?"

"Yes, lieutenant this is Airmen Devon calling with the PPC out at Randolph, we have a change of assignment for you which requires fast action on your part."

"What exactly are you referring to?"

"Well sir there was a delay in your change of duty station orders, you will be receiving your hard copy by COB tomorrow. You need to be on station by 1500 hours this coming Monday, and we know it is very short notice. Your family members will be sent for after your arrival."

"Well, I have no family that needs to move; where is this new assignment taking me?" Harper began to think, *damn I'm going to miss this place, and I've seen a lot of interesting stuff taking place.*

"Nellis AFB, Nevada."

"Nellis."

"Yes sir, Nellis, location details will be provided in the hard copy."

"Ok, I understand, well not completely but this is the Air Force."

With most of his fellow team members completing their calls and reporting back to Sergeant Reed. Reed was compiling the report that Devon would need to give to the Colonel.

Reed was at Devon's cubical.

"Are you about done?"

Hanging up the phone.

“Yes, that was my last one, I was able to make contact with everyone except for one individual that was on scheduled leave.”

“Who is the missing person?”

“A lieutenant Johns assigned to the 14th FTW.”

“Ok. Have you called his commander yet?”

“Yes, but all I got was some clerk, I left a message for them to return my call ASAP.”

“Call to the lieutenant’s commander’s office again, stress to whoever answers the phone the importance of the matter.”

Just as Devon was about to pick up the phone it began to ring.

“Good afternoon, Airmen Devon speaking.”

“This is Colonel Sharpe returning your call.”

“Yes sir, we have a Redball action on one of your assigned personnel. We need to have them located and advised of their new assignment. They will need to be at their new duty station at 1500 hours this coming Monday. Also, we need them to return a call so that we have a verbal commitment and understanding from them. A hard copy of the orders will be in your office by COB tomorrow.”

“Why Hell, what's with the short notice?”

“All I know is there was some kind of delay in the system and now it’s all out of whack.”

“Ok, I’ll get some people to try and locate him.”

“Thank you, Colonel.”

Devon went over to inform his section chief.

"I just spoke with Colonel Sharpe he is getting some people to try and locate the one person that I was not able to reach."

"Well, that's good news. Ok, you can leave for the day. First thing in the morning you need to start processing the orders and get them transmitted to all the commands. Follow up with a call that they have received them, we don't need any more errors."

Devon was walking out to his vehicle when Jackson called out to him.

"What's up, Mike?"

"So how did you like your trip to the commander's office?"

"Oh damn, my heart skipped about three beats. You're not going to believe this crap. Some arrogant jerk was like ripping me a new one earlier on the phone, then telling me don't move from my desk that he was going to have the higher approval call me within ten minutes."

Mike just started laughing.

"Hell, it's not funny, this guy rips me, then the Colonel rips me, and then Reed was like pulling my chain about being late from my break."

By this time Mike was falling all over the place laughing his butt off, at Devon's expense.

Finally catching his breath, he was able to ask another question.

"So, what was the issue all about?"

"I'm not too sure, but whatever it is, it must be good."

"Whatever do you mean?"

"Well, the system was keeping the orders in the pending file, as needing higher approval."

"Yeah, that happens all of the time."

"Well like I said this guy called about it and not even ten minutes later the colonel had me in his office about it. Then tells me the President called him about it."

"The President?"

"Yeah, that's what I was saying, the President. Now that I think about it, I've been processing orders for Star Search for about six months. It's almost like they are assigning a whole new base. I may not be a rocket scientist, but I know when something is going down, and I'm telling you, something is going down.

Mike and Devon decided to continue their conversion over at the NCO Club it was time for a cold one.

Sergeant Reed gave the end-of-day report to the colonel, of which most was good news. The team was able to contact all but three of the personnel on the list.

"Tell me more about the three that have not been contacted?" Asked the colonel.

"Two of them are in the middle of an ORI (*Operational Readiness Inspection*) their commander has been informed and will provide any needed assistance. We are expecting to hear from them early tomorrow, the other one is the individual who was out processed. We are waiting on legal to give us a determination on his recall status."

"Well, it seems you have a good handle on this situation, better than I was expecting."

"Thank you, sir; it was a team effort."

"Ok, let's meet up after lunch tomorrow unless there are any surprises I need to know about sooner."

"Yes, sir."

Just as the day was winding down at the Personnel Center it was starting to heat up at the Palmdale plant and Groom Lake.

February 27th, 2008, Palmdale Ca., 2030 hours

Maxwell, along with Hodge's, and Chavez's team met at the designated time.

Hodge started the short meeting informing them that they were ready for the live fire and that they would be going over to the engine test cell building.

The team members were assembling outside and boarding a bus. Pretty much everyone was quiet, there was not much talk going on.

It was a short ride over to the test cell, only about seven minutes. You could tell that you were at the right location by the number of M.P.'s stationed near the building.

Upon getting the authorization to enter the team went inside, and there she was the first completed bird from Palmdale. Located in the building next to this one you would find bird number two.

These two live-fire runs would be just like the ones Maxwell did a few days ago back at Groom Lake. The only difference this time was Maxwell would just go through all the flight control checks but would not do any taxing of the aircraft.

Upon the first craft getting a complete thumbs up from Maxwell you could see a small sign of relief from Hodge and his team members.

The time was approaching 2215, and with everyone moving over to the second test cell, Hodge presented a question.

"Seeing that it's going to take about another hour would you guys like to postpone this to tomorrow?"

Hodge asked, looking towards Maxwell and Chavez.

"I would rather get it done tonight, as I have a strong feeling that I will be needed back at Groom Lake." Informed Maxwell.

With the second test finishing up near midnight, it was getting very late. With all members back on the bus, over to the main building, they went. Soon they would retrieve their vehicles and go back to their lodging facilities to retire for the night.

"How about we meet up at say 0900." Hodge inquired to the group.

A few heads motioned up and down in acceptance of the proposed time.

What seemed like only minutes was hours, six hours to be precise. Buzzzzzzz was the only sound that Maxwell heard in his dream; wakening up and realized it was the alarm clock and not a dream.

The same process was being had in the other rooms of Chavez, Tony, Jack, Rex, and Louie.

Lynda was already down in the lobby going to get some breakfast.

Once more everyone had gathered in the conference room of the production building.

Hodge once more took charge of the meeting.

“Well, last night was just fabulous as you all know. There was a message for you Maxwell to call Colonel Starkes ASAP. Also, GL has advised that there would be more engines on the way.”

As the team met for a few more minutes going over the last few day's work, Maxwell left the room to go call Starkes.

“Yes, sir this is Maxwell.”

“How are things shaping up there?”

“Great, we completed full systems check on the first two birds last night with the newly installed engine cores.”

“That’s good to hear, there are more engine cores on the way; also, we have two C-5s that will be on station today to transport both of those birds to Groom Lake. The fighters from Nellis will be providing an escort, operations should have the orders and notify the crews this morning. I need for you to return today; Rogers has the computer ready for us to do systems check on the UE1.”

“Yes, sir I will be ready to leave by 1200 local.”

“Great.”

Chapter 6

February 29th, 2008, Groom Lake, 2100 hours

The fortress of a hanger that was housing the UE1 was in a full beehive mode the entire day. All-section teams and department heads were meeting, going over every possible detail. No one was wanting to have any errors or mishaps under their direct control.

"Ok Rogers, it's all in your hands." Informed Starkes.

Once looking over the checklist that he had prepared, Rogers took a long hard look around at the Bridge. Pressing a switch on the console enabled him to speak to the whole ship.

"Ok team, turn on your recording devices."

Each team lead began to turn on the video recorders that were given to them to document their section.

"Allegra, would you confirm that the hanger is clear and sealed?"

"Yes, sir." Picking up her headset and positioned it on her head.

"Ground control do you copy?"

"We copy, loud and clear."

"We need a confirmation that the hanger is clear and sealed."

"Stand by UE1."

"Control to all stations, report status."

If one were to investigate the massive hanger the picture would be quite different than what they saw earlier in the day. There was not a single soul moving about.

"Station 1 clear and locked."

"Station Two clear."

"Three clear."

"Four clear."

The report kept coming in clear and locked to station seventeen.

After a few short minutes.

"Control to UE1."

"UE1 copy."

"The hanger has been checked and all doors are in lockdown mode."

"Copy control."

"Lieutenant we are clear and locked."

Taking a deep breath, Rogers steadied his nerves. *"Ok, here we go."* Thinking to himself. Speaking into his headset. "SMART turn on the internal auxiliary power."

"Auxiliary power on," SMART replied with an automated response voice.

As soon as the power reached its prime load, several of the stations on the bridge began to power up, along with the one in the engineering section.

Looking over to Maxwell. "Captain, how is your console looking?"

"As far as I know it looks good, the lights are on."

"Communications is yours on?"

"Yes, the auxiliary light is now on."

"Evo?"

"Yes, we are on."

"Pressing a button on the console, labeled engineering.

"Chief is your station on?"

"Affirmative."

"Colonel we are hot across the board." Stated Rogers.

"Preceded lieutenant, it's your show right now."

"Comm, have them kill the external power."

"Yes, sir."

"UE1 to ground control."

"Go for ground."

"Kill external power."

"Copy, kill power."

The ground controller moved over to a large breaker box located on the wall and pulled the switch down to the off position.

"Ground to UE1, we are in the off position."

"Copy ground. Lieutenant the external power is off."

Looking over to the indicator light for the external power source Rogers was able to see that it was indeed out. They were now standing alone on their power capabilities.

Calling back down to engineering. "How are the readings chief?"

"Per the indicators, we are reading normal on the core drives and the aux is still on the minimal side, she can take on more power demands."

"Evo, fire up your systems."

"Yes, sir."

Jackson reached over to the primary power switch, lifted the safety toggle, and flipped the switch."

Lights started to flicker, along with a few short beeps. Deep down into the internal workings of the ship a slight hum noise was heard as the onboard oxygen system finally came to life. As Jackson was monitoring the various gauges and controls you could see one for the o2 sensor reading at ten percent. It would take a few minutes for the system to be at one hundred percent capacity.

Looking over to the other readouts, the onboard sensors were reading steadily; oxygen twenty-one percent, Nitrogen seventy-eight percent, Argon one percent, and carbon dioxide at point zero three eight percent. Otherwise known as air, it was reading at one hundred percent.

For the past few years, all the air that had been breathed on the ship was being pumped in from the external connections. Along with the air in the ship being filtered through air scrubbers on the roof of the hanger.

The light changed from red to green and the o2 sensor was now reading one hundred percent, which means the ship was now able to fully produce a breathable air supply.

Located throughout this ship in some key strategic areas, are some smaller air recirculates which would filter out toxins in the air and remix it with the incoming air feed.

Jackson looked over his console board and all lights were showing green; all the readouts were in their normal position. UE1 was able to sustain itself, as far as air goes.

"Rogers the Evo is green across the board and ready to stand alone."

"Great, thanks."

"Comm, have ground kill the air feed."

"Yes, sir."

"UE1 to ground."

"Go for ground."

"Disconnect the air feed and close off the scrubbers."

"Copy stand by."

Just as soon as the order came in the ground crew technician went over to the wall and pulled the switch for the air feed. Next to it was one labeled air scrubbers, throwing that one off as well.

Even before the ground crew was able to call back in Jackson already saw the lights change on his console, the external connections were off.

"Rogers we are totally on the ship's air system," Jackson reported.

"Ground to UE1, you are disconnected."

"Copy ground."

Allegra confirmed what Jackson had just informed Maxwell about.

"SMART run a systems diagnostic on the AUX and EVO," Maxwell spoke into his headset.

It would not be very long before the system check was complete, Rogers was thinking.

"AUX diagnostic checks out complete one error located in grid 10.521.55, EVO check complete. There are ten errors on record in section 5.5, 5.6, 5.7, 5.9, 5.10, 6.22, 6.23, 6.24, 6.25, 6.30.

There appeared to be a high look of alarm among everyone present; more like dazed and confused.

"Does anyone know where those sections are located?" Starkes yelled out.

Everyone began to talk over each other.

"Hold it, hold it, everyone shut up." Starkes once more yelled out.

"Rogers?" Said Starkes.

"Yes, sir." Speaking once more into the headset. "SMART where are those error sections located?"

"Port Hanger Bay."

"Ok ok, this is why we are here. Rogers continue with the test."

"EVO, what are the Air Pressure readings?"

"Fourteen point seven on the bridge. The same for Engineering and Medical. Starboard is reading Fourteen point six, Port is Fourteen point Two and dropping, forward is Fourteen point Five along with Aft."

"Security seal Port." Rogers directed.

The Security chief pressed a switch, which would close the blast doors leading from the Port hanger to the main body of the ship.

"It's closed."

"EVO report any change." Directed Rogers.

"NAV fire up your system."

The Navigation officer reached out to turn on the guidance system.

"NAV is on, but we are not able to lock on for a location."

"That's ok, I was expecting that the optical scanners are not able to see any of the stars while we are in the hanger." So said, Rogers.

"Ok, the ship's pressure is at fourteen point seven in all sections except the Port Hanger, which is now at Thirteen point six and still dropping."

As the night went on there were many small issues that would need to be addressed but figuring out what was going on in the Port Hanger was going to be the major priority for the team.

"Rogers, can you have a debrief report ready for 1100 hours tomorrow?" Starkes inquired.

"Yes, sir that will be attainable."

"Great, let's all meet in the Auditorium at 1100."

One by one each section chief went through the shutdown procedures for their section; just as the ship came to life now it was going dark, dark in the sense as it was all quiet and most of the console lights had gone out.

Overall, the fire-up inspection was an enormous success; there still would be many more items to check. But it would seem, that all is on track for the launch date this upcoming summer.

Rogers was in his living quarters going over SMARTS reports. Feverishly looking at the data trying to put together a mission report and recommended solutions.

March 1st, 2008, Groom Lake, 1100 hours

With all the section chiefs present along with some of their selected team members, everyone was ready for last night's UE1 testing debrief.

As the Colonel entered the Auditorium; someone called out. "All rise."

Immediately Starkes responded. “Be seated.”

Taking up center stage Starkes began to inform the team what they already knew.

“Last night was a huge milestone for this project; given the circumstances; the fire-up test was a tremendous success. Rogers would you take the podium.”

Starkes stepped aside and took up a seat near the left side of the stage. Rogers standing and moving into position. Looking out into the small crowd of people it was starting to hit home. Having to speak to this group, most of which were, at least, a good seven years or so older than himself was not going to be easy.

Opening the attaché case Rogers removed his notes.

“The main issue, and our biggest area of concern, as most of you know is the loss of pressure of the Portside Hanger. Once the Blast door was closed which sealed the Port hanger from the rest of the ship the air pressure was maintained at fourteen point seven throughout the vessel. However, the Port Hanger was unable to maintain proper pressure. We will need to cover every square inch to locate the leak or leaks as there could be more than one. The best way will be to do a pressurization leak detection using a color-activating gas to find out the cause and location or locations of the air loss. There were many other codes identified, most of which show various sensor readings outside of their given parameters. After a complete system diagnostic, each component should be changed out. I have compiled a list for each department to run a diagnostic and subsystem check for the system faults. They're roughly a hundred and twenty-three items to cover. Medical along with the Hydroponics section had no reported issues to address. Are there any questions?"

The room was quiet, no one raised a hand. Starkes stepped back to the podium.

"Ok, we need a team lead from each department to come and receive a copy of their section report. Return a timetable to the coordinator by 1600 today on your expected completion date. If there are more than five items map out a detailed flow of completion. If there are no questions, you are dismissed."

One by one each of the team leads came forward, it was not long before all the assignments had been passed out and the room was once, became empty. Only Starkes, Rogers, Elizabeth, and Maxwell were left.

"Great work guys, very splendid work. Maxwell, I'm going to need you to fly me up to Nellis on Monday, plan for a 1200-hour departure. Elizabeth, plan to start the next orientation class Wednesday. I will be meeting with a group Monday, and they will be on station by COB Tuesday. Rogers, I need for you to work closely with the coordinator so we have a plan of attack and a timeline in place, I would think that we will be able to re-test in about Two weeks."

Chapter 7

March 3rd, 2008, Nellis AFB, Nevada, 1500 hours

Starkes and Maxwell were standing backstage in the makeshift auditorium. Which had been placed in the building. Most of the seats had been filled, there were only a few left empty.

The two of them began to make their way to center stage.

"Room Ten Hut." Was called out by one of the persons sitting up front.

The room became quiet as everyone stood up. Starkes took to the podium.

"Be seated."

Just as fast as they stood up, all the individuals took to their seats.

"Good afternoon, first I would like to thank all of you for getting here on such short notice. Due to the nature of this project, it was unavoidable, that it had to take place. To answer a question that I'm sure most of you are asking yourself. Why are you here? Well, ladies and gentlemen, I have been tasked with developing a special needs project. Every one of you has been selected to fill a role on this team. The first order of business I need each person that is married to please stand."

Looking out into the group of personnel about forty people were standing.

"I need for you to all come forward and follow Captain Maxwell, he has a separate briefing to conduct with you."

Maxwell, lead the small group off into a smaller room.

"Ok, the reason you are here. First, you are bound by your security agreements if for some odd reason you don't think that you will be able to live up to those standards I ask that you leave this room now!"

After a short pause, with nobody getting up to leave.

"Great, with a show of hands how many people here feel that there is some other life form in existence in the universe?"

Once more looking over the group, Starkes saw about fifty percent of the hands were raised.

"For each of you that stay in this room after this next question, all I can say is your life is going to drastically change. Change in what you do, where you live, and the way you see life in general. If you are not able to go on a no-contact assignment that will last for an undetermined amount of time, you are asked to leave now with no consequences to your career.

Speaking from under his breath Harper could be heard by the person sitting next to him. "Oh damn, I don't believe it."

"What don't you believe?"

"Nothing, but I know both of them quite well."

"So what's up, you staying?"

"Hell yeah; you?"

"I don't know, undetermined sounds like a long time."

"Well if it's what I think it is, you probably would want to stay."

"What makes you say that?"

"You raised your hand to the question about other life."

Once more Starkes addressed the room. "Ok so one person left, this is your last chance to leave. Any takers?"

After a short pause, with no one else getting up.

"Great, I have been the commander of the UE1 project for the past few years, this is a project that is located at a classified location as of now you will not be able to have any contact with anyone that is not part of this team. Is that clear?"

The room in unison answered with an astounding yes.

"Harper would you please stand."

As Harper stood up the guy next to him looked in astonishment. *Didn't see that one coming he thought.*

"Harper please tell the group where your last duty station was located."

"Sir, I'm more than sure you know that information is classified."

"Yes Harper, I am, but that is where everyone in this room is going. You are authorized to inform them of the location."

"Yes, sir. As you know I was stationed at Groom Lake, Nevada, or as some like to call it Area 51."

Suddenly there were whispers and small talk taking place throughout the group.

"Keep it down. Yes, you are now assigned to the UE1 project located at Groom Lake, each one of you has been selected for your expertise in your career field. What you don't know is that the UE1 project is going to be the world's largest space vehicle. We will be departing this planet in a few months, hence the undetermined amount of time that we will be gone."

Hands started going up, there were questions to be asked.

"Let me finish, then I'll answer a few questions. Yes, little green men are real, well so far they are gray but very much are real. The vast majority of the advances in our aircraft and imparticular this ship are driven from the alien craft that crashed in Roswell back in the fifties, along with a few others throughout the world. What needs to be implemented by all of you is the ability to keep an open mind. There is too much in the known and unknown universe which we don't know anything about. Questions?"

Hands went up.

Starkes pointed to one person in the first row.

"You said we leave in a few months. Is that date set?"

"Yes, the launch is set for July." Starkes points to another person.

"What will our living quarters be like?"

"On base, you will be in multi-unit buildings two to a room, on UE1 depending on your rank you will be assigned to either an individual room or a quad unit."

"Where are going?"

"That's a great question. We have our sights set on a star called Sigma Draconis known as Alsafi it is about 18.8 light-years from our sun. We are to explore the Habitable Zone of that star."

Starkes called out and pointed to one of the members on the back row.

"You said the intended location is 18.8 light-years?"

"Yes, that is correct."

"Ok, I'm not sure but doesn't the speed of light move like 180,000 miles per second."

"Close, it's 186,282 miles per second. I think I know where you're going with this question. We are anticipating arriving there in about twenty years or less."

Upon that last statement, there was a lot of mumbling and chatter throughout the room. Several people began to turn and talk to the person next to them.

"Any more questions?"

Once more hands went up as the noise became quieter.

"Yes mam, on the third row."

"I know that I'm single but someday I was hoping to have children. By the time we return, if we do return a large part of my life will be over."

"That is an excellent statement, this mission has many interesting parts. One of the many aspects is the ability to sustain life, which is for us to reproduce; the married personnel will be allowed to bring along their families. Single people like yourself will be allowed to marry and start a family if they so chose. With around 5,000 personnel assigned, and two-thirds of them single, there is a large pool of people for relationships to develop from."

Now more hands were raised.

Pointing to a young man in the second row.

"What if we already have a girlfriend? I've been together with her for several years."

"There is a possibility for some to be allowed, there is a complete vetting process that first has to take place. Several items are the overall needs of this mission. Questions that need to be asked would they be able to contribute in a way that would be beneficial to the program? We have many needs, but the first people brought on are to fill our critical needs. Part of your orientation process is your family and personal background and the possibility that they might be able to fill some less critical needed roles."

As the questions kept coming in, in the main auditorium, Maxwell was with the smaller group.

"Ok, ladies and gentlemen the reason you have been separated from the rest of the group is your needs are slightly different from the others. As each of you is well aware the Air Force can PCS you without your family. Each one of you needs to ask yourself these questions. Can you be away from them for a very long period? Will they be willing to come along? All of you have a spouse that will fit into the program. You have been selected to fill a critical role. This mission comes first. But there is a place for each of your spouses to become a valuable team member."

A few hands went up.

"I know this raises some questions, but bear with me for a minute. Who in here believes in some kind of Extertresial Life?"

Most of the people in the room raised their hands.

"It's good to see that most of you do. This is going to be the first time that you will hear an official statement on this matter. Alien life forms are real and they have been to our planet several times over the years."

"This next question will undoubtedly change your life as you know it. It is not going to be unrealistic that you will be able to return anytime soon. Knowing that your immediate family will be allowed to go with you, who will be able to commit for say twenty years or so?"

There was not a sound to be heard, you could only hear a slight noise from the air blowing through the vents.

"This is your only chance to leave without any recourse, once I proceed with this briefing you and your family will be committed to this program."

Two people got up to leave, with one person hesitating, then stopping, turning around.

"My wife is pregnant with our first child, she is due in a few months."

"All I can tell you is we are aware of that and we have an excellent child care facility available."

"What if she doesn't like it?"

"If you stay and this goes for all of you, you will be committed and so will your spouse. You are speaking for them as well."

"This is a once-in-a-lifetime event, you have all been selected out of all branches of military and civilian employees of the government. More than likely you will never see the likes of this ever again in your lifetime."

After taking a few more seconds to ponder and let his brain digest what was said, Anthony went back to his chair.

"Now that we have that out of the way let me first remind you all that this is held to the utmost level of security. I would like to welcome you to the UE1 team. What is UE1? UE1 is short for Universal Explore One, our first true spaceship. One that will take us far beyond the moon, and the planets in our solar system. We are going to Sigma Draconis or some call it Alsafi, it is 18.8 light-years from earth. So for those asking a light-year is ruffly about 186,000 miles per second, in short, very fast. For our mission to fully succeed we need married people who have or willing to have children. There is something for everyone on board, it is a completely self-contained city, with movies, playground areas, and various other forms of entertainment for adults and children. Now I'm sure a few of you might have some questions?"

No sooner had Maxwell got that last statement out when hands went flying up. As it was with the group that stayed with the Colonel most of the questions were the same.

"Ok, let's go rejoin the other group."

Maxwell took them back to the main auditorium.

"Well, it's good to see that most of you have stayed with us. Now that we are all back together it is about time for us to leave. Maxwell check with Flight Op's to see if they are ready for us."

"Yes, sir."

Maxwell stepped to the backstage area to make a call to the Flight Ops.

"Nellis Flight Op's Airmen Knight speaking, how may I help you today?"

"Yes, this is Captain Maxwell checking on the status of our transport."

"Yes sir, do you have a sortie number?"

"Yes, it is 1411 code name Box Car."

"Ok, let me check on that for you."

After a few minutes. Airmen Knight got back on the line.

"Sir the status shows to be ready for personnel arrival."

"Great let them know we will be there shortly."

Walking back to where Starkes was still speaking to the group. Maxwell informed the Colonel.

"Sir they are ready."

"Great!"

The Colonel readdressed the group. "Ok, we are going to head out to the back. There should be two buses that will take you all out to the Flightline. We should have a couple of C-17s waiting to transport you all to your new base of operations."

Just as soon as the Colonel finished his last statement the group began to exit the building, outside just as he had said two buses were waiting for them.

The drive out to the waiting aircraft was a very short one, it took longer for everyone to get situated and settled into a seat than it did to drive out to the tarmac.

"Maxwell, file a flight plan for Palmdale, I would like to leave at 1600. I want to check in on Chavez and his team."

"Yes, sir."

Maxwell dialed the number to the flight op's center.

"Nellis Flight Op's Airmen Knight speaking, how may I help you today?"

"This is Captain Maxwell again, I need to have my Bird prepped for departure ASAP, destination Palmdale."

"Sir, we will need to have orders sent down."

"Just type in Codename Shadow, all of the authorization you will need will be there."

Knight typed the codename into the required field on his computer screen, the screen changed to a flashing red border around the edge. Displayed in the middle of the screen were the words in bold green letters, *Approve, and expedite any requests.* Airmen Knight had never seen this on his screen before in the field, only once back when he was in tech school.

"Yes, sir we will have your bird ready for you in an hour."

"Great, just call me back at this number when it's ready."

"Yes, sir."

Knight called out to one of his coworkers. "Hey Lorraine, remember back in tech school when they were going over the flash messages on the terminal?"

"Yeah, why?"

"I just got a 'Redball Action'."

"Are you for real?"

"Yes, come take a look."

Lorraine went over to look at the screen.

"Damn, it is real, I thought that was just some B.S. the instructor was pulling on us back in school. We better make some calls, fast!"

Picking up the phone, Knight calls over to the Base Transient Office.

"BTO, A1C Reed."

"This is Knight over at base Op's, we have a Redball Action for Captain Maxwell, Tail 550, destination Palmdale, ETA 1 hour."

"Roger, Knight, 55 zero DT Palmdale 1 hour."

Meanwhile, Maxwell had asked the Colonel if they wanted to grab a bite to eat before they leave.

Sitting at a table in the Bx Food Court, A young individual approached them.

"Excuse me, Sir, my name is Dean Ford, and I was one of the people in a briefing that you had a short while ago, but I choose to leave."

Starkes looked at the young man. "What is it you need?"

"Well sir, I was thinking and if it is possible I would like to join, whatever it is you have going on."

"Well son, you missed a lot of information, and this is not a place to discuss this matter. But if you are sure."

Taking out a piece of paper and writing a few words on it, hands it to Ford.

"Take this to base op's and they will advise you on what to do."

"Yes, sir. And thank you."

"No need to thank me, you will have time for that later."

As Ford leaves, Maxwell asks.

"So what was his specialty?"

"A cook."

Maxwell laughs.

"A cook?"

"Yeah, we all got to eat. Besides his evals have all been superior. And speaking of eating, we need to do so too, we got a plane to catch."

It was not very long before Maxwell and Starkes had consumed their meal, and were just about ready to leave when Maxwell's phone rang.

"Hello."

"Yes, sir, this is Knight over at base op's your bird is ready and you can go to the Transient office whenever you are ready sir."

"Great, thank you."

It was a short drive over to the Transient office, parking in front of the building the parking space was marked for Offical Government Vehicles Only.

Upon entering, they were greeted by A1C Reed.

"Good afternoon, what can I do for you today sir?"

"You should have a bird ready for us, and our flight gear is in the locker room."

“Yes it is ready, and you may go change if you are ready to leave. Do you have a vehicle?”

“Yes, here are the keys.” Maxwell hands them to Reed.

Not much more is said as the pair walk down to the locker room to get changed into their flight suits.

Once they have changed and picked up their go bags, they proceeded out the back door. Outside there were a few blue trucks parked, all had markings on them. *Nellis Base Transient*

As they walked up to the first truck.

“Sir, which aircraft are you going to today?”

Asked one of the guys standing by the first truck.

“55 zero,” Maxwell replied.

“Ok, we have been expecting you. If you like you can place your bag in the back and we will head over to your aircraft.”

At about the same time, Ford was entering the base op’s center. Walking up to the counter he met up with Knight.

“Hello, what can I do for you today?”

Ford hands the note to Knight. Looking at it, he reads. *Code Name BOXCAR 2 DT* 1134.15

Knight types in the code name, and once more he sees a flashing red border and in green bold letters in the middle of the screen Approve and expedite. Towards the bottom of the screen, it reads enter DT. The cursor is flashing waiting for Knight to enter more commands.

“Lorraine, Lorraine, you are not going to believe this!”

"What is it now?"

"We got another Redball Action."

As she is heading toward him she blurts out.

"No bloody hell, in two years I've never seen one and now we got two in one day!"

"Well, what is the DT?"

"I don't know, there are just some numbers on the paper he gave me."

"Type it in let's see what happens."

Knight starts to type in the numbers, 01134.15. Pausing and looking at Lorraine.

"Go ahead, push enter."

Knight presses the enter key. The screen flashes bright red and displays the word Alert three times in the middle of the screen.

As Knight and Lorraine look at each other, somewhat puzzled. Ford asks.

"What is it?"

Lorraine answers. "I don't know, I've never seen this before."

Then out of the blue, the office phone starts to ring, but it is different, there is no pause it is just one long continuous ring. Picking up the phone, Lorraine answers.

"Base Op's A1C Smith speaking."

"Smith this is Sec Def command, who is in charge?"

"I guess that would be me, sir."

"Ok, someone there just inputted some information into the computer which triggered an alert action, is that right?"

"Yes, sir my coworker."

Ford and Knight are just listening and both of them looked very confused.

"On what authority did they act on?"

"It was a note from a guy who just walked into the Op's Center."

"Put them on the phone."

Lorraine looks to Ford. "They want to speak to you." As she hands him the phone.

"Hello."

"Who are you?"

Ford replied, stuttering, "I'm Senior Airmen Ford sir."

"Ford, I don't need to ask you where you got that note from. What I need to know is are you ready to leave?"

"Leave sir, to where?"

"You will find out where soon enough. Do you need to get your bags?"

"Just what I have in the VAQ, I just got here earlier today."

"Great, here is what's going to take place." As the Sec Def office was speaking two MP's entered base op's.

"You will be escorted by MP's to clear out of VAQ, they will bring you to the base alert hanger. At that time you will then be placed on a transport plane to take you to your new duty assignment. Do you understand?"

"Yes sir, is that going to be right now?"

"As we are speaking it is all being placed into motion. Now it's time to go."

As Ford gave the phone back to Lorraine, you could hear it over the radio from one of the MP's.

"Base to unit 12."

"Unit 12, over."

Escort Sr. Airmen Ford to VAQ, and expedite to Alert 6, over."

"Unit 12 copy, over."

Ford turns around.

"Are you Ford?"

"Yes."

"Let's move."

The two MP's practically dragged ford right out of the building and took off at a high rate of speed with the lights and siren blaring.

Knight, turn's to Lorraine. "What in the hell was that all about?"

"I don't have a damn clue, but have you ever heard of the Secretary of Defense Office calling?"

"No."

"Well, something big is going down, two alert actions in one day."

It took less than a minute to reach the VAQ. Parking the car right up to the entrance of the building.

"You have one minute to get your belongings, then we are dragging you out."

"One minute, I need to check out and pack my clothes."

"I don't think you understand, just throw your junk in the suitcase or bag or whatever it is you have. There is no packing, we don't have time to waste, your butt is getting on a plane in less than three minutes, with or without your belongings."

"Ok, ok, I get your drift."

Ford runs into the building with the MP's right behind him. One MP drops off to speak to the desk clerk, the other stays with Ford. Down the hall to the right they go, luckily Ford's room was on the first floor.

Once Ford made it to his room he was fumbling with the room access card, dropping it. Finally, he had the door open stepping into the room there was not much there just a duffel bag and two suitcases. Most of what he had brought was still in them, just a few clothes were thrown on the bed and some personal items in the restroom.

Back out to the patrol car, bags tossed into the back seat, and once more off like a bat out of hell.

"Don't worry you are checked out, so you don't have to worry about any more charges on your account."

"Ah, thanks, I guess."

You could tell that Ford was very puzzled, the MP's did not show any concern at all, just another day on the job, orders, just more orders to follow for them.

Just as they had said a few short minutes ago, they pulled up to the alert aircraft hanger. Jumping out of the car each one grabbed a bag and trotted over to a C-130, which already had its engines running, the props were just buzzing making a whoosh sound as they turned around. Being greeted at the entry ladder to the aircraft, Ford got his bags from the MP'S.

Sergeant Jones introduced himself. "I'm Jones, you must be our package."

"I guess I am, Sr. Airmen Ford."

"Ok, we need to go it's wheels up in one."

"Wheels up?"

"That's right, now up the ladder you go."

No sooner than Ford and Jones were in the plane Jones called out to the aircrew.

"Package is on sir."

"Roger that, close and secure."

Just as soon as the pilot had said those words the plane started to move.

Jones closed the hatch and showed Ford where to sit.

"Nellis tower Alert 114 on the move."

"Alert 114 you are clear for immediate departure, all ground activity is holding in place, go around if need be."

"Copy tower."

"So Jones where are we going?"

"I have no idea, I thought you might know."

"Hell no, all I did was take a note to the flight ops center and bam, here I am five minutes later."

You could hear Jones laugh a little bit under his breath, but you heard him.

The C-130 began to increase speed, the engines started to make a much louder whoosh and hum-type noise. And there goes the nose of the aircraft starting to lift off of the runway, then the rear of the plane, followed by the landing gear being retracted.

"Nellis tower A114, airborne, over."

"Copy, A114, take angles 33° North, maintain 8 zero and pick up 000.00 on your VHF, over."

"Why would they have us go to 000.00, there is nothing there."

"A114 copy, we are holding at 8 zero and squawking to 000.00, over."

"Who knows, key up the mike and see what we get."

As the pilot turned to the new coordinates, the co-pilot called in on the radio.

"A114, over."

"A114, this is A51, you are to maintain your current heading you are clear for 11N, outer marker 223 is reading winds out of the South at zero 8 pick up the slope at 223 over."

"A114, A51 clear on 11N maintain heading, marker 223 wind zero 8 out of South on the slope at 223 copy."

"A51, A114 you are clear all the way. You are about 10 minutes out. Take the first taxiway, pick up MPs to ramp over."

"A114, A51 first taxiway MPs to ramp over."

"Well, if that doesn't take the cake, first there is someone there on a radio frequency that doesn't exist directing us to a runway that's not there."

"Check the flight map again."

"For what? How many times have we seen that thing, there is nothing there but a huge empty piece of dirt."

The locator beacon began to flash, 223 showed on the digital screen.

"Well there goes the beacon, what else are we going to find?"

"Damn look at that."

"What do you make of that?" Replied the co-pilot.

"Ok, this is starting to make sense, now."

"What's making sense?"

"Duh A51, it's the infamous Area-51."

"Oh, hell."

It was not much longer and they were on the ground following the MP truck to a parking spot on the ramp.

Chapter 8

March 4th, 2008, Palmdale Ca., 0730 hours

Chavez was patiently waiting for the colonel down in the lobby of their hotel quarters. Walking down the hall, Chavez sees Maxwell heading his way.

"Maxwell, so why is the colonel here?"

"He did not say much about it, we were out at Nellis and he told me to make a flight plan to come out here. I think he just wanted to put some eye's on the prize if you catch my drift."

"Well, that makes me feel a bit better. I got a really bad case of the butterflies in my gut."

"No need to sweat it, from what I gather everything is looking good."

About that time Starkes was making his way down the hall to the lobby.

"Good morning sir, how was your flight."

"The flight was fine but the in-flight meal was atrocious." Smiling at Maxwell when he said it.

"Well let's go get some grub, so we can talk."

"Do you want to eat over at the plant or somewhere else?"

"The plant should be fine, I hear they have a pretty good cafeteria."

"It is excellent sir, there are a lot of choices and I don't recall having a bad meal yet."

It was a short drive over to the plant, and Chavez was driving.

"So how is the production line moving along?" Starkes was asking Chavez.

"From what I have seen, it looks to be moving very efficiently."

"Have they given a production timeline and the estimated delivery of the aircraft?"

"It has been brought up and they feel very confident that they will be able to meet the schedule that has been requested. The main item holding them up are the engines and fuel cells that are being produced at GL."

"Yes, I have been thinking about that and we are just going to transport them without the engine cores. We will install them at Groom Lake, which should cut the timetable down by a few weeks. Besides, it will help the maintenance troops get their training and efficiency up."

Pulling into a parking spot, Chavez was feeling more at ease.

"We have to get you checked in and get your entry badge."

"Then let's get a move on, I'm getting pretty hungry."

It did not take too long to get Id badges for Starkes and Maxwell. Off to the cafeteria, they went.

Chavez directed them to his favorite table in the place. The conversation continued, discussing in more detail, the required fixes that Chavez and his team were sent up to work on.

"What time is the daily production meeting?"

"It is set for 0900 hours, we still have some time."

"So how is this guy, what's his name?"

"Roy Hodges, he seems like a pretty straight arrow. I must say he has been very helpful, and willing to work with us in any way possible."

"Good, then he should not mind too much the news I have for him."

"Would you like to enlighten us, colonel?" Maxwell asked.

"In due time, in due time. So how about we head to that production meeting Chavez."

"Yes, sir it's back upstairs on the first level."

Entering the conference room Roy was already there along with the rest of Chavez's team.

"Good morning all, Roy I would like to introduce you to our commander, Colonel Starkes."

Roy moved over to greet the colonel, extending out his hand.

"Well, it's very nice to meet you, sir."

"And you as well. I've heard some good things about you and your team here."

"That's is reassuring, as you know we always strive to produce the best product possible."

"So what is the current update?"

"Well sir, with the engine cores and cells that were brought in we will have them installed by the end of the week. We should be up to full production very soon, by the middle of the month at the latest."

"That is great to hear. That will make this latest bit of information easier to take in."

"And what might that be?"

"We need to have twenty more X33s ready by June twenty-fifth."

"Twenty more, and earlier than planned. I just don't see how that's going to be possible."

“Well, here's something that will help. We will be sending you some more workers most of them have been working on the ones that are at Groom lake. So they are up to speed, also, you will not need to install the engines or the fuel cells. We will complete that after they have been sent to Groom Lake. We need to expedite the process and have the fleet ready and flight tested. We need to get our aircrews certified as well. They will be going operational in July.”

“July, that is a huge undertaking. How do you ever expect that to happen in such a short time?”

“The task is doable. With all of the bugs worked out now, and with the successful test flight and engine runs. We feel it is achievable. The formal contracts will be sent down from the contracting office by the end of the week. My understanding is there is a spectacular on-time delivery bonus that your executives are going to be very excited to see.”

“Well then, I guess we have our work cut out for us. How would you like to see the production floor now?”

“Why yes, that would be great if you could show me around some.”

“Ok, guys does anyone have any questions or concerns that need to be addressed?” Chavez asked.

Everyone was just shaking their heads no.

“Well then, let's get to it.”

And with that, everyone departed except for Roy, Maxwell, and Starkes.

“Maxwell, get with command and check on the status of the transport. Make sure they will have a couple here on Friday so that we can get the completed birds back to the base.”

“Yes, sir.”

“Roy, I understand this is not what you were expecting to hear today, but we have the utmost confidence in your company’s capability to deliver the assets on time.”

“Well thank you for that confidence colonel. So if I might inquire, what is the rush? Are you afraid we are not going to be here?”

“It’s something like that, but no you will probably be here just not us. So how about that tour.”

“Just follow me, I see that you have a visitor badge, so we should be all set.”

Roy began to lead the way out and down the hall to the access elevators. Maxwell stayed in the conference room to make some calls back to the base of operations to check on the status of the transports. As Maxwell picked up the phone he began to ponder the question. *I wonder how long he plans on staying here?*

March 4th, 2008, White House, 1130 hours

“Mr. President, the Russian ambassador is here sir.”

“Good, send him in. Notify me as soon as Sec Def gets here.”

“Yes, Mr. President.”

Yuri Ushakov enters the oval office less than ten seconds after the aid leaves the room.

"It's good to see you, Yuri, how have you been?"

"I'm doing fine Mr. President, it's been some time now."

"Cut the hogwash, I've told you to call me George."

"Yes, sir George it is. I'm just not used to the informalities, George."

"There that's better. See that was not that hard was it?"

"No, it was not."

"Ok, then I hope that you have some good news for me?"

"Well sir, it would seem that the ISS (International Space Station) is moving along well. Now that your country has installed Node 2 we will be able to complete the docking station for this UE1 craft, and we are ready to launch back-to-back Soyuz missions to take up the sections that we agreed to."

The President's intercom buzzes.

"Yes, Karen."

"Mr. Gates is here."

"Great send him right in."

As soon as Gates enters the room and the door is closed, all three men shake hands.

"Robert, I believe you have met Yuri Ushakov."

"Yes sir, it's been several months."

"Well, we were just talking, and Yuri has reassured me that the Russians are ready to deploy back-to-back missions. So where do we stand with the UE1, General?"

"From the last night's progress debrief the project is on schedule. It might even be a few days ahead, now that they have completed the power run-up testing with the onboard systems. It was noted that there were a lot fewer discrepancies or potential problems than what was initially planned for."

"So when will the first section be ready to send over the pond."

"According to Starkes, they should be doing another power-up in about two weeks. If the issues are cleared up they will begin the sectional removal of the pods by the end of the month. The first transports should be flying out in early April."

"I sure hope so. The Russians are not going to be able to repeat these shadow games. We are only going to get one shot at this."

"We will have the assets in Anchorage in time to be loaded on the Antonov An-124, one on the way to the Airshow in Canada, and another on its way back."

"Ok, so that will get two of the pods to Russia. So how is the third one going to get there?"

This one gets a little bit more tricky. In May there will be a joint military exercise with NATO. The Icelandic Air Policing. It will be common to see a large number of fighters and support aircraft there in support of this operation. The Russians will have one of their heavy transports nearby declare an emergency, which will allow it to land. In the middle of the night, we will switch over the pod to their aircraft. The next day they will leave under fighter escort, then they will return to the Cosmodrome in Kazakhstan. Where there just happens to be a Soyuz Rocket launch set for the ISS."

"Well, this doesn't sound like there is much room for error."

"No sir, there is no room at all."

"So, Yuri, are you going to be able to pull this off?"

"We will make it do. We have a lot riding on this just as you do. We have given you more Titanium in the last four years than we have been able to produce in the last twenty."

"I know that your government has been very diligent in delivering the required amount of raw materials that we have needed. Your government has done an excellent job of keeping it a secret, but we are not talking about a few hundred pounds at a time now. We are talking about success or failure and the latter is not an option that either one of our governments can except. Now is it, Yuri?" Said the President.

"No, it is not, but what is it your people say. Keep calm and carry on."

"Ok, then I think we have a mutual understanding. How about some lunch?"

"That would be great."

Meetings with the Russian ambassador could always be intense and there was always a hint of mistrust, but General Gates was getting used to it.

President Bush escorted Yuri and General Gates out of the oval office and down to the private dining room. As was customary when the president was in the White House, lunch would be served promptly at 11:45. There was no exception today. Other than a few high-profile guests would be in attendance, which was not out of the question.

March 4th, 2008, Palmdale Ca., 0947 hours

After A brief tour of the facilities, Starkes met up with Maxwell back in the conference room.

“Notify flight ops, we are going to Edwards AFB. I would like to leave say about 1330 that will give us some time to gather our belongings and have lunch with Chaves and his team.”

“Yes sir, I will make it so,” Maxwell replied to the colonel.

Maxwell moved over to the phone. Called the operations desk, once more to get their bird prepped. It would be a short flight to Edwards, but just like all Military flights the aircraft would be checked and re-checked.

Maxwell was thinking, Edwards? *I wonder what is going on at Edwards?*

Edwards AFB was the prime landing facility for the Space Shuttle Program, back in the early days of the program. There is a very large dry lake bed, you had miles of available runway to use.

After speaking with Base Op's Maxwell called for a base taxi to take them back to the hotel.

Not much was spoken on the way to the hotel. Starkes seemed to be off in a daze as if he were somewhere else, anywhere but in sunny Palmdale California.

Once back at the hotel Starkes advised Maxwell that they would leave at 1130 hours. That way they could meet up with Chavez and the others to have lunch and discuss the upcoming influx of workers that would be arriving in the next few days.

Maxwell went to his room, it would only take five minutes or so to pack his go-bag. Thinking this would be an excellent time to take a nap, well second thought, how about a call to Elizabeth.

Maxwell placed a call to Nellis AFB, most calls were routed through Nellis for security reasons.

Finally hearing the phone ringing, Maxwell waited anxiously. Needing to speak with the woman that he was now totally in love with. After all of these years and now it happens to him out in the middle of nowhere, love.

"Training, Sargent Riddley speaking."

"Riddley, it's Maxwell, is Elizabeth available?"

"One second let me check."

After a brief period of silence.

"Maxwell, I'm surprised to hear from you. How are you?"

"I'm great, wishing that I could be with you."

"Well, you will soon. How's Nellis?"

"Nellis was ok, but it was short, we are in Palmdale. We are fixing to leave heading out to Edwards."

"What happened at Nellis? Palmdale, now Edwards I'm having a hard time keeping up with you."

"Not much, Nellis pretty much went as planned. But the colonel wanted to meet with Chavez and the X-33 team. Now just out of the blue, it's off to Edwards. And I don't have a clue as to why."

"Well, that makes two of us. You just be careful, I worry about you when you are out and about."

"Well thank you for the concern, but I'm a very well-trained aviator you know."

"That's not the point. It's just that I care for you."

"The feelings are mutual. You are on the top of my radar screen too, pun intended."

"I'm being serious here, and you are making jokes."

"No, but it is my job to fly you know, and I've been well trained by the USAF to do just that."

"Well just be careful, call me when you can."

"I will, just as soon as it's possible."

Still over an hour before he was to meet with the colonel, Maxwell thought it would be wise to take that nap. Time to rest before they set off to Edwards.

It seems that an hour is getting shorter and shorter each day. Before Maxwell knew it, it was time to meet up with the colonel.

Starkes was already down in the lobby by the time Maxwell arrived.

"So what's with the change of plans to go to Edwards?"

"Who said it was a change of plans!"

"Well you never said anything about going there, I just thought it was a last-minute change."

"No, more like just need to check on some items that have been sent over to be placed onto the next shuttle missions."

"So what do the shuttle missions have to do with our project?"

"You might say they have everything to do with our project, we have been sending components to the ISS for a few years now. Some very critical items will be sent up this month and the next few to come before our departure."

One must agree that the colonel is a master of dishing out crumbs and keeping oneself on their toes so to speak.

Starkes and Maxwell arrived at the flight operations center, checking in with the duty officer. They were directed to wait in the lounge area.

Not long after getting situated one of the transient members came in for them.

"Captin Maxwell." He called out.

"Yes, that's me." He replied as he began to stand up and greet the Airmen.

"If you will follow me, sir, I have something to show you."

"Is the aircraft ready?"

"That's what we need to show you, sir."

As they walked out to a waiting truck the two of them took a short drive down the flight line to where the jet was parked.

Walking up to the jet, the Airmen directed him to the centerline external fuel tank.

"The seal started leaking after it was filled. We have called for a fuel truck to remove the fuel. It's going to be at least a twenty-minute delay. Then we can drop the tank and see about fixing it."

"Just remove the fuel, they can look at it when we get back. Our destination is not that far and we will have plenty with the wing tanks."

Back over at the operations center lounge, Starkes was questioning Maxwell.

"It will be fine sir, as long as we are not going cross country we don't need that much fuel."

"How long are we looking at before we can leave?"

"They said a twenty-minute delay."

"Well, that's not too long it could be worse."

Sitting down in one of the lounge chairs, Maxwell was viewing the television that was in the room. Starkes was on the Sat phone that he always carried with him.

Not much time had passed when the Airmen came back for them.

"Sir we are ready for you."

Once they got over to the aircraft and placed the bags into the travel pod. The transient crew closed the cover and locked it down with the fasteners. Starkes proceeded to climb into the back as he was becoming very familiar with it. Maxwell did his walk around and looked over the maintenance logbook. Once he was satisfied that all was in order up the ladder he went.

Sitting in the cockpit, going through the checklist Maxwell was checking the switches. Made sure all was in the right position before he called into the tower.

"Tower 55 zero we are ready for engine start."

"55 Zero you are clear for engine start."

With clearance given Maxwell spoke into the headset after he changed the radio setting to the ground communications channel.

"We are clear for engine start."

"All clear sir." Responded one of the transient personnel.

After going through the flight control check and calling in for clearance to taxi, Maxwell and Starkes were once more on the move. Soon they would be airborne.

It was a very short flight over to Edwards. Once more they were parked in the transient area of the flight line. Even with their delay back at Palmdale, it was still early in the day. Just a hair past 1430 hours.

"Welcome to Edwards sir, how long will you be staying?"

Starkes was just getting down the ladder.

"We don't have a definitive timetable set, but I would expect it to be a few days." Advised Starkes.

Thinking to himself Maxwell began to wonder why they are to be here for a few days. A few days seems like it would be three or more days.

"So Colonel would you care to clue me in, as to why we are here?"

"Sure, you need aviators if you are going to command a flying wing. This is where you find them."

"How is that sir?"

"Did you ever hear of the movie *The Right Stuff*?"

"Yes sir, it's one of my favorite movies."

"Then you should understand, this is where they are, flight test school."

Maxwell and Starkes put their bags in the transient truck and took a seat in the back of the bed.

"Over to VQ?" Asked the driver.

"Take us to the 412th Operations Group," Starkes replied.

"Yes, sir."

After a short few minutes of drive, they were in front of the building. It did not take much time for them to depart from the truck and enter the operations group building.

Once inside the entrance lobby, you saw a cleric at a desk station.

"Good afternoon sir. What assistance may I provide for you today?" Asked the airman at the station.

"We are here to see the Colonel."

"Are you referring to Colonel Panarisi?"

"That would be correct." Answered Starkes.

"Do you have an appointment? And who may I say is calling?"

"Colonel Starkes, with training command, and no we are not expected."

Starkes and Maxwell located some seats in the lobby and sat down to wait for the commander, as the cleric called down to announce their presence.

"Sir there is a Colonel Starkes and another officer here with training command to see you."

"I don't recall anyone from training on my calendar for today."

"No sir they are not on your calendar."

"Ok, thank you, Jaz."

After about a ten-minute wait Captain Jones entered the lobby reception area.

"Colonel Starkes."

Starkes and Maxwell both stood up to greet Captain Jones.

"The Colonel is in a meeting. What can I assist you with today?"

There are not many things that will set Starkes off but getting dismissed by someone he wishes to speak with and getting the errand boy. Now that is just one of those pet peeves that will do it.

"Well son, the first item is you can direct us to the Colonel's office or better yet just escort us there now. Whatever he is doing can wait for what we need to discuss."

The captain may be of lower rank but he is not going to be talked down to in his duty station.

"Colonel, I'm not sure where you are from but this is our house and if our commander is busy then he is busy. I would suggest you schedule an appointment and return at that time."

"I tell you what, you go crawl back under whatever rock you just came from that way when the crap hits the fan in about five minutes you don't get too much stink on you."

"Sir I do not and will not take that attitude and threats from you. I'm going to have to ask you to leave, now!"

"I don't think so, we will just have a seat, I'm pretty sure the Colonel will be here soon enough."

And with that Starkes sat back down, to be followed by Maxwell taking his seat as well.

Jones went over to Jaz. "Call over to the M.P.'s to have these guys removed from here."

"Yes sir."

By this time Starkes was already on the phone, calling the SecDef.

"Good afternoon Robert, sorry to bother you so late in the day. I got a tad bit of a problem here at Edwards."

"What is it? Too much sunshine for you?"

"No, we are trying to meet with the commander of the 412th OG and we seem to be getting brushed off."

"No worries Jerome you will have your audience with Panarisi in less than two shacks of a dog's tail."

It did not take very long for Colonel Panarisi's phone to ring. Colonel Panarisi knew by the line that the call was coming in had to be someone important. This was a back-office line for him and not many people had this number.

"Colonel Panarisi speaking."

"Panarisi, this is SecDef. You have a visitor in your lobby that needs your attention. I suggest you get your butt down there and give them whatever it is they need. Do you understand Colonel?"

"Yes sir."

And just as fast as the call came in the line went dead.

Panarisi started to head over to his door, to go meet with whoever this Starkes was in his lobby.

It did not take very long for two M.P.'s to show up. Jones met them at the door and was explaining the situation to them. In a not-so-nice tone of voice, they directed their look and movement to Starkes and Maxwell.

“We need both of you to come with us, now!” Stated one of the MP’s.

Starkes was not phased at all by them. However, the look on Maxwell was quite different.

“I don’t think so. We have a meeting with Colonel Panarisi. And I’m very sure he will be here soon, no I’m positive he will be here in a few seconds. So no, we will just wait here.”

“Sir I’m not going to ask you again. The next time we will remove you by force.”

“You just try it, sonny, it will be the last mistake you make in your short military career. I will have you knocked down to Airmen Basic and sweeping the sidewalk with a toothbrush.”

The M.P who was doing all of the talking was just a three-striper, and here is a Colonel not budging one bit. Keying the mic on his radio.

“Charlie three to base.”

“Go Charlie, three.”

“I have an officer that is not complying. Can you send a supervisor asap, over?”

“Copy Charlie three. S two is en route.”

As Colonel Panarisi is walking down the hall he is hearing all of this commotion take place. He picks up his pace, almost running by the time when he makes it to the lobby.

“Oh crap, what the hell?” Panarisi speaks out.

“It's ok sir, they will be leaving very soon.”

"No, no, there is just a big misunderstanding. I believe we have a meeting, and you are?" Panarisi is saying as he is walking up to Starkes.

Now Starkes stands up along with Maxwell.

"I tried to tell them that you would be here very soon. Colonel Starkes and this is Captain Maxwell. I assume you have somewhere we can speak in private."

"Yes, we can go to my office if you would just follow me please."

As they began to walk off Jones's mouth dropped almost to the floor. Looking at Panarisi.

"I'll explain later Jones."

As the three of them started walking off the M.P.'s were looking just as confused as Jones.

"Thank you for responding so quick but it would seem that it was just a big misunderstanding," Jones advised the M.P.'s.

After a brief walk down the hall and into the Colonel's office.

"Would either of you like something to drink?"

Maxwell spoke first. "A bottle of water would be great."

"The same for me." Replied Starkes.

Panarisi picked up his phone and after a short pause asked for some water bottles to be brought to his office. Less than a minute later there were two firm knocks on the door. Panarisi stepped to the door. Jaz met the colonel with three bottles of water.

"Thank you, hold all of my calls for now."

"Yes, sir." She replied.

With all three of them sitting across from each other, Maxwell and Starkes on one side and Panarisi on the other side.

"Well, let's cut the bull out and explain to me, how or better yet, why the SecDef just called to advise me that you are here to see me?"

"Now that's my kind of officer straight to the punch, well Mike, if that's ok by you. We are here to recruit some of your top aviators. We need pilots that are highly trained and able to adapt to a changing environment, in location as well as their equipment."

"You got to be kidding me, you are manning some kind of special op's group and you're coming here to steal my men."

"No, we are recruiting the United States government's personnel. Due to the training and testing that has taken place here at Edwards over the years, this is where we are to find some of the most capable aviators on the planet. Now if you like I could make another call and this time it's not going to be the SecDef on the line. I'll just get your commander in chief for you."

"The president, you're going to call the president."

Looking at Panarisi, Starkes has a huge machination-eating grin on his face and trying to hold back the laugh that is building inside of himself. Starkes pulls out his phone. Pressing the number one which is set for a speed dial to the president.

"Why Jerome I was not expecting to hear from you for a few days. Is everything alright with our project?"

"Yes sir, we are looking great. I just stopped off at Edwards to meet with the commander of the 412th OG. Looking to complete our personnel needs, it would seem he is a little reluctant to let go of some personnel."

"Oh, I see, so I need to speak to him?"

"I think that would be a fabulous idea, sir."

Handing the phone to Panarisi. Maxwell was now sitting with a huge grin on his face.

"This is Colonel Panarisi and you're are?"

"Why Colonel this is your Commander in Chief, President Bush. Now before you go and question the fact that you are speaking with me you will be getting a priority message from your aid in a few minutes. Now that we have the pleasantries out of the way you will listen and comply with whatever Jerome asks of you is that clear?"

"Yes, Mr. President."

"Now put Jerome back on."

Handing the phone back to Starkes the grin on Panarisi was completely gone.

"Yes George, I take you had a pleasant conversation?"

"If you need anything else you know where to find me."

Just about the time Starkes had put his phone away, there was a knock on the door. Panarisi got up to open the door and there once more was Jaz.

"I know you said no interruptions sir but this priority message just came in for you sir."

Taking the envelope. “Thank you, Jaz that will be all.”

Closing the door and returning to his seat.

“So what exactly is it you need from me?”

“Why don’t you need to open that message?”

“No, I'm pretty sure it is just a confirmation to extend any courtesies to you. And assist with any requests that you might have. I don’t’ know what the hell is going on but I will concede that you have the power to execute it.”

“Colonel, our work is on an extreme need-to-know basis. Of which you don’t need to know the full details as of now. Although I will enlighten you a bit, we are with a unit attached for now to Groom Lake. We are needing to fill some aerospace technicians positions along with top-quality aviators. The best and brightest work their way through here. Our mission is going to require that they go voluntarily without full disclosure as to where and when they will return. We need to have a commander's call to take place tomorrow say about 08:30 hours should work, then we will need to review the selected personnel files for those individuals that raise their hands and seek to move forward with our program. After we select the people we need, they will PCS on Friday. We will have a transport plane here for their immediate departure. Now unless you have some questions I believe you need to make some calls it get this setup. So if you don’t mind would you please have someone call for a base cab so we can get checked into the VQ and grab a meal, I’m getting tired and hungry.”

Panarisi looks as if he just got his teeth kicked in, but gathering his composure and moving over to his desk phone calls to Jaz.

"Jaz call for a base taxi and have Captain Jones report to my office."

As Starkes and Maxwell began to see themselves out, Starkes looked back at Panarisi.

"Colonel we are not here to twaddle on your parade. In fact, with us being here it is going to look very favorably for you and your program. Have a good day sir."

Back out in the hall as they preceded to the front of the building Jones passed them in the hall, still puzzled he was.

Just about to knock on the Colonel's door Panarisi looked up.

"Come in Jones and close the door."

Panarisi was still sitting at his desk. Jones took up a position in front of his desk standing at attention.

"Just take a seat."

"Yes, sir."

No sooner had Jones sat down than Panarisi started to explain the events that had just taken place and the event that needs to take place.

"Zero 830 sir, that's short notice. With it being this late in the day, it's going to be extremely hard for the units to communicate with all of their personnel."

"Just inform each unit commander to use whatever means they need to get the message out and have their personal in attendance."

Chapter 9

March 5th, 2008, Edwards AFB, 0825 hours

A base cab just pulled up to the VQ, Starkes, and Maxwell walked outside to get in.

"Good morning, sirs. Where to today?"

"Building 1600."

"1600 it is."

"Well sir, it looks as if most of the group is here. So where are our mystery men?"

"That would be a good question Jones, but one for which I don't have an answer. Go see if you can find them outside."

"Yes, sir."

Looking out into the crowd of personnel in the hanger most had taken a seat. There were a few groups scattered around the hanger taking and trying to figure out why they were all here on such short notice.

With the time approaching 0835 the base cab pulled up to the hanger. Starkes and Maxwell exited the vehicle, upon approaching the entryway they were met by Captain Jones.

"Good morning Colonel, if you would please come with me we are ready to start."

Starkes and Maxwell took up a stride behind Jones who was leading them to the backstage area that was set up for the commander's call.

Jones keyed his two-way radio. "Alpha Two to Alpha One, over."

"Go for Alpha One."

"The guests have arrived and will be in position, over."

"Copy Alpha Two."

Colonel Panarisi motioned to one of his aids.

"Luitenant Quill moved closer to the colonel.

"Yes, Sir?"

"Call the room."

"Yes, sir."

Quill went out onto the stage to the mic.

"Room ten hut."

Instantly the hanger became very quiet everyone stood up and the small groups all turned to face the stage. Standing tall and straight at attention.

Now with the room quiet and ready to receive the words from the unit commander. Panarisi was moving to the center of the stage where the podium was located.

"At ease, you may be seated. First I would like to thank you all for making this on such short notice. We would like to extend a warm welcome to Colonel Starkes. He will be speaking with you on a very important matter so without further ado, Colonel, the floor is yours."

Panarisi stepped to the side of the podium as Starkes and Maxwell walked out.

"Thank you Colonel Panarisi. Good morning and thank you for your attendance. A question that I'm sure most of you are asking is why you are here? Well, ladies and gentlemen, I'm going to answer that for you right now. I'm the commander of a special project developing an ultra-secret and very unique aerospace vehicle. For this, we need a large personnel group to maintain and fly the craft. So we are here to fill positions in all specialties from the ground personnel to the aviators that will fly it. Now I know you have been working with numerous types of vehicles here and testing new and inventive ways of their use, but this is something very few people have seen, and even less have flown. Only one individual has had that experience so far."

Looking over to Maxwell.

"To my right is Captain Maxwell he will be meeting separately with those of you who are pilots. I will be meeting with the rest of you who stay for phase two of this brief. First I need to remind everyone in this room that you are bound by your security agreements if for some odd reason you don't think that you will be able to live up to those standards I ask that you leave this room now!"

After a short pause with nobody getting up to leave.

"Great, with a show of hands how many people here feel that there is some other life form in existence in the universe?"

Once more looking over the group Starkes saw a lot of hands go up.

"For each of you that stay in this room after this next question, all I can say is your life is going to drastically change. Change in what you do, where you live, and the way you see life in general. If you are not able to go on a no-contact assignment that will last for an undetermined amount of time, you are asked to leave now with no consequences."

Looking out into the hanger what started as a group of about three hundred and fifty was now reduced to about two hundred.

"Ok, I ask that everyone move up closer to the front."

Slowly the group filled in the space that was vacated when the others left with the back few rows empty.

"Our next step is to review your current work assignment and skill set, unfortunately, not all of you will be selected for this assignment. We only have so many open slots but I would like to thank you for your courage in taking this step. I know you don't completely know what the work assignment is yet but for some of you, you will soon enough know."

Starkes looked toward Maxwell which meant it was his time to speak.

"I need for the pilots to please form a line to the left of the stage, all of the ground personnel to the right. We will be passing out a short questionnaire there are no right or wrong answers, but you need to be completely honest in your answers. Only put something down if you believe in it. Don't put something you think we want to hear, it will make a difference. After you have filled out the questionnaire you may leave, we will be notifying those of you selected no later than 1300 hours on Thursday. Phase two brief will take place at 1600 hours at a location to be determined."

The room was separated down the middle into the two requested lines. One by one they moved forward getting a questionnaire form. The forms were similar but different. One by one each person returned the form filled out and left the hanger.

The time was fast approaching 0930 hours by the time the last form was returned. With them, collected the task at hand was to review them for the mission needs and review their personnel file.

Starkes looked over to Colonel Panarisi.

“I assume you have a small conference or work area that we can use for the next day and a half?

We also would like your input as well, so we will need your direct assistance.”

“Yes, we have a conference room back over in our headquarters building. I also would suggest that Jones be included as he has more direct contact with the unit commanders and personnel.”

“That would be fine. So if we can catch a ride with one of you. We need to get moving there is a lot to cover and we are on a tight deadline.”

Walking out to the parking area, Starkes was telling Maxwell to have Two C-17s on standby for Friday.

“We have Two C-17s assigned here at Edwards.” Informed Panarisi.

“Well, that’s convenient, cancel that request then Maxwell. Colonel if you would see that they are ready and available for a Friday departure, 1400 hours to Nellis AFB.”

Back at the OG building they were getting set up in the conference room.

“What can I do to help?” Asked Jones.

They had the questionnaires placed into two stacks. Start with the ones from the pilots, and separate them into two piles. One for married and the other for single. I will start with the other stack.” Said Maxwell.

Looking to Colonel Panarisi. “Colonel Panarisi if you will get a couple of your admin personnel we will have some names for them to start pulling their evaluation folders. Also, I need an office to make some calls." Stated Starkes.

"You can use my office," Panarisi said as he got up to go get the requested admin personnel.

It was only a few minutes before the two stacks were now four. Maxwell handed Jones a template.

"Take one of your stacks and match this up if all of the blocks are filled in, put it in one stack if something is missing then put it into another pile. So you will have two new piles one married and matched and one married and not matched. Do the same for the single personnel."

By the time Panarisi returned with some admin personnel, there were eight piles on the table. Pilots married matched and unmatched, pilots single matched, and unmatched and the same for the ground personnel from Maxwell's stack.

"This is Jaz who you already met and Lisa they will pull the files for you."

"Jaz take the pile that is married pilots matched and Lisa you take the ones that are single and matched." Maxwell requested as he handed each of them a stack of papers.

As the two women left the room Maxwell directed Jones to make two more piles. "Take the married and unmatched if there are three or more missing put them in one pile and in the other pile we only have two or fewer missing.

By the time the ladies returned there were stacks all over the table. Placing the pulled files on a separate table at the end of the room they now were left with questionnaire sheets for the ground personnel, married and matched and single and matched. After a short time, they had eight stacks of files on the table. From here they would make their selections. Counting up the sheets that had three or more missing blocks they just cut 52 people from the 212 questionnaires they had when they started. It only took about 45 minutes to weed out the first twenty-five percent but the next cut was going to take somewhat longer.

"Ok, let's start with the pilots, take the married stack if they are under thirty place them in one pile and the rest into another. Anyone over forty a separate pile."

Having the pilots separated Maxwell had Jones start on the ground personnel, while he started reviewing the folders. The first ones he looked at were over forty. For the most part, the pilots needed to be as young as possible. For a trip that was going to be over twenty years one way. What was this person going to do, still be flying at sixty years old? And if they were to be on the return trip they would be about ninety years old. The same would hold true for all of the personnel, but the stress of flying a fighter aerospace vehicle was just not the same as one who did not fly.

Jones completed separating the other folders.

"Ok, what would you like for me to do now."

"Take a look at the names and files from here."

Pointing to a new stack Maxwell had been making.

"See if anything pops out at you that might not be in the official file that we need to know."

About this time Starkes and Panarisi rejoined them in the conference room.

"How is it looking?"

"It's progressing well, we had 52 cuts from the first round and about 15 more so far. But I don't have a firm count." Replied Maxwell.

"Good, what do you need us to do?"

"If you would start reviewing that stack and Colonel Panarisi if you would look over the ones Jones has gone through we need to know if anything stands out to you that might not be in the official file. We don't care what it is. There is to be no impact on their record but we must have the right fit for our program."

Starkes looked up at the time it was going on 1300 hours and was getting somewhat hungry.

"How about we take a short break and get a bite to eat at the officers club?"

"I could go for some lunch, but we don't have an o club anymore. It is now Club Muroc and it is open to all of the personnel. There is a dining area in which it is pretty quiet." Informed Panarisi.

"Good enough, gentlemen if you will lead the way."

Club Muroc was not busy as it was mostly past the lunch hour. Overall the meal was short but the nourishment for their body was greatly needed.

Returning to the files it took the rest of the day to complete the review and to separate the individuals that would get the offer. They all were very qualified, but for some, it might have just been their age. Others it was some of their beliefs. Two got disqualified for their prior attitude remarks in the personnel file. All in all, 78 got scratched for one reason or another.

“Let's call it a day. Colonel if you could have some admin support personnel at 0830 we will get the individuals notified. Where can we get a group of 134 people together that’s not too large of an area and one that can be made secure?” Starkes asked.

“We have an auditorium in this building that will work.” Replied Panarisi.

“Great then we shall see you at 0830.”

Starkes and Maxwell called for a base taxi to take them to the VQ. But first, they decided to hit the food court at the Base Exchange.

“Popeye’s Chicken sounds like a winner tonight.” Suggested Maxwell.

“I think I’ll have Arby’s.”

As each one waited in their respective lines, it was about five minutes before they met at the table to enjoy one more deluxe fast food meal.

Once the meal was done Maxwell informed the Colonel that he needed to pick up some items in the BX.

"Sounds like a great idea, I'm sure there are a few items I need as well."

Once back in his room at the VQ Maxwell knew it would be a good time to give a call to Elizabeth.

"So how long are you going to be gone?" She asked.

"I'm not sure but I think we might be coming back Friday or Saturday, we have another batch of team members that are to be transported on Friday. After that, we should be done here, so hopefully, we will return then."

"That would be great, I feel like it's been weeks since you were here."

"I know, I feel the same way."

"Well it's getting late and I need to get a few hours of rest, hope to see you soon"

"Ok, my love, sleep well."

Maxwell was the first to arrive in the lobby of the VQ, waiting for the Colonel. Maxwell called for a base taxi.

As the Colonel approached Maxwell, he was still on the phone putting in the request.

Hanging up the courtesy phone. "Good morning sir."

"Yes it is, it's a very good morning."

"You seem to be in very good spirits today sir."

"Why yes, I do believe that is so. If most of our recruits elect to join the program then it looks like we will have all of the people that we need to fill the designated slots."

"So does that mean we will be going home
tomorrow?"

"As of now, that would be an affirmative yes."

Now that was some great news to hear. Maxwell was looking forward to returning to Groom Lake.

The ride over to the 412th seemed very short today but now it was time to complete the magic and get the newest members on board.

Panarisi brought in three clerics to assist with the calling. And with Maxwell and Jones that made five that would be smiling and dialing the morning away. This would mean that each person would need to make about 25 calls each. Maxwell figured they would be done before the lunch hour.

After all of the calls were done and a good lunch break Maxwell and the Colonel went to the meeting area to prep and go over the disclosure. Starkes would cover the first half and then they would break up into two groups. One would be the pilots with Maxwell. Starkes would take the rest of them.

About an hour before the designated meeting time, Maxwell decided to take a walk outside, just kind of looking things over.

1545 hours, time to get back inside thought Maxwell. Upon entering the meeting room there was already a fairly large crowd present.

Two M.P.'s were posted at the door. Not yet keeping anyone out or asking for any Id. One would wonder why they were here.

"Maxwell, let's get them seated, and close the room down."

"Yes, sir."

Maxwell went to the entry doors where the M.P's stood. Requesting that the room be secure no one is to enter. Once back inside Maxwell called out in a much louder voice than usual. "Room ten hut."

Just like the other day the room became instantly quiet and everyone was now standing facing Starkes.

"Take your seats," Starkes commanded.

Handing a stack of papers to Maxwell. Maxwell began to distribute them out to each person in the front row. Starkes informed them to take one and pass it back.

"Does everyone have a form?" No one raised their hand or spoke out to indicate that they did not have one.

"Good, read and sign. As you do I will give you the highlights of the document. Each one of you is a member of the armed forces of the United States. You are about to be made aware of information that has been designated above Top Secret. You are expressly forbidden to repeat any information that you receive today in this room with anyone other than a member who is now presently located in this room."

"First order of business we need to vet each member when you're name is called stand and answer present. Make sure you say it so it can be heard."

Maxwell began to read the names off of the roster that Starkes and himself compiled earlier in the day. One by one each person stood. One hundred and thirty-four names were called and all one hundred and thirty-four individuals were standing.

"Very good, you may be seated. Outside of this room, there are very few people who know the complete story and all of the events that have taken place over the past fifty-plus years. Most of you in this room have been directly involved with or very closely worked with some very unique and state-of-the-art aircraft. Take a good look at some of them, dim the lights, please."

As the room became darker, Starkes started the slideshow. Appearing on the screen was a photo of the SR-71 with a date of 1966, followed by the B-2, 1989, F-117 Nighthawk 1981, Space Shuttle 1981, U-2 Dragon Lady 1957.

"The list can go on and on. You should start to get the picture. Now here is one you have not seen and I'm guessing you have not even heard of it."

On the screen is a static photo of the XY-33.

"This is the newest line of defense, lights please."

Someone in the back turned on the lights as Starkes kept talking.

"For the most part, everyone who is a part of this team has done so on a volunteer basis. What I'm about to ask you is very unorthodox, for the nature and operations that you have been accustomed to. It is a crucial requirement and need for the vital success of the task we will be undertaking. First, for those of you that are married, you are going to have to answer this question with them in mind. Also, any of you that have children, you are answering for them as well. For the rest of you that are single, it may be easier to answer. There will be no recourse if you choose to pass on this special opportunity. If you choose to move forward you will be on a means of transport at 1600 tomorrow. For those of you who are married or have a family, they will be brought out within about two weeks. There is one other special craft that I have not shown you but it will be your home."

A lot has been said, but when you think about it, still there has not been much told at all.

"Lastly, there is a very good chance those of you who choose to accept will never return to this planet."

Bam there it was the bombshell that everyone was waiting to hear. Hands started going up, expressions were changing, and the intensity and excitement in the room just went into overdrive.

"Yes, we have developed a spacecraft which will allow us to leave our know nine planet universe and travel to a planet which is about 19 light-years away. We are hoping to get there in about twenty Earth years. To answer some questions I know you have. Yes, those of you that are married your spouse will be allowed to come, along with your children. We have accommodations for all of their needs. Those of you that are single and have a significant other in your life, they also might be able to come. It is on a mission-needed basis, there is no guarantee that we will be able to accept them into the program. The civilian members and children are needed for the long-term success of this project and for others to follow in our footsteps."

Starkes was letting more out than what he usually would but he knew this crowd was different. Besides, he needed all of them to join the program.

"Is there anyone who would like to leave before I complete the briefing? We are at the point of no return."

No one stood up, no one left the room, and best of all no one was freaking out. It looks like the questionnaire did the job.

"Ok this is your last chance if you stay you are committing yourself and if you are attached you are speaking for them as well. There is no turning back once you agree."

Still, no one left.

"All right then. You might be wondering how did we come up with all of this technology? Well for the first time you are going to hear from an official government employee. Aliens are real, there are other life forms out amongst the stars and it's time for us to expand our horizons and accept the truth. So who has questions?"

Hands were up all over the room. Pointing to the first row. "Yes."

"Are there plans to return?"

"Excellent question. Yes, we do plan a return trip although the ship is designed with individual living quarters for the married personnel and four-person bay pods for the unattached. Some individuals will be left behind if we find a habitable planet. That also will be on a volunteer basis."

Pointing to a lady in the second row. "Yes, mam."

"I put off having children for my military career. I still would like to have some one day in the future. If this is a twenty-year trip one way I'm going to be about sixty-five years old when we return."

"Part of the success of this mission rests on us reproducing, so single people will be allowed to marry along the way. It is highly encouraged for them to do so. Right now I think we are about a 35 percent mix of personnel who are married. Leaving 65 percent of the crew that is single. Our crew size of about 5,000 people leaves a large portion available, about 3200 people who will be looking to mingle."

Looking to the back of the room Starkes motions to a guy near the door. "You sir"

"So when do we leave?"

"Tomorrow, you will depart at 1600 from here. A C-17, which will take you to Groom Lake."

Starkes was pointing to a guy on the third row who had his hand up.

"So what does this UE1 look like?"

"I think you are the first person to ask me that question. It is the largest ship that you have ever seen. It is larger than an aircraft carrier and completely self-supporting. I can talk about it all day but seeing is going to be the best way to understand the size and magnitude of the vessel."

So if there are no more questions, then you need to report back here by 1530 tomorrow. You will need to have a basic go bag with a few changes of clothes and your essential personal hygiene items. Take this time to get personal affairs in order as it is very unlikely that you will be returning here anytime soon."

Chapter 10

March 12th, 2008, Groom Lake, 0915 hours

Maxwell was at the entry point to the main auditorium greeting the pilots as they were entering. Directing them to the front rows. All in all, there were fifty-two pilots accepted into the program. Most of them would be assigned a dual role. Meaning they would have other functions besides flying. It was going to be a long journey and who knew when they would get to fly the XY-33 once the UE1 got underway.

Maxwell began to walk down the steps to address the group. As Maxwell stepped up onto the stage area someone called the room to order.

"Ten-hut."

All became quiet as each person stood.

"At ease, take your seats," Maxwell instructed.

"Squadron good morning."

Most if not all of the assigned personnel responded with.

"Good morning sir."

"Well, now that you have gone through your orientation and you have the basic feel for the UE1 we are going to start on the XY-33. Are there any questions?"

"A few hands went up.

"Yes, Rivera."

"Is it true what we have heard about the speed, sir?"

"Well I don't know what it is you have heard, but I will just say that when we did the first test flight it only took a matter of seconds to reach low orbit. About 75 miles high, along with putting NORAD on high alert."

This last part of the statement drew some laughter from the room. Johnson had his hand raised.

"Yes, Johnson?"

"How does it feel compared to other aircraft you have flown?"

"Well, there is no comparison. The speed, maneuverability, and the response from your thoughts to the output of the vehicle are undeniable the most incredible thing I have ever experienced."

Davis now had his hand up.

"So when do we get to see it?"

"Now that probably is the million-dollar question. First, we will be reviewing the tablets that were assigned to you during your orientation. After you complete the general knowledge and get the basic understanding down we are going to move to computer simulation. Let me tell you the simulation is almost as good as the real thing. Once you have gotten some hours in we will start with an engine warm-up and then proceed into some nighttime taxi events. Then if you are up for it, we will start the flying phase. So who is ready to cover the flight control icon on your tablet?"

What seems like a long process, was one that would have to be completed in a very short time. With it already being mid-March, they would only have about three months to complete all of the required steps. Along with assisting in the engine testing as the completed XY-33s were delivered and learn their additional work assignment.

As the morning was progressing along, the group was beginning to get a better understanding of the aerospace vehicle. Covering the design basics, flight controls, and the basic principle of the engines. As the time was nearing the lunch hour Maxwell dismissed them till 1330 hours.

"We will meet back in the classrooms after lunch Maxwell instructed.

As they were leaving for their lunch break Elizabeth was making her way down to meet with Maxwell.

"Well, mister instructor would you like to take me out to lunch?"

Maxwell was pondering to come back with a smart-aleck response but decided to take the safer road.

“Sounds like a fabulous idea mam.”

As they turned to go back up the stairs, one of the pilots was making his way back down.

“Excuse me sir, but I have another question that really can't wait.”

“Yes, what is it?”

“Well, sir I find myself attracted to Martinez and I was wondering if it would be ok to ask her out?”

Both Elizabeth and Maxwell began to laugh.

“Sir I’m being serious.”

“We understand, we are laughing as we also are dating. Having met here as well. So your answer is yes. But you must know that our mission comes first above all.”

“Understood sir.”

“Well it would seem as thou love is once more in the air,” Elizabeth stated.

“It would appear so. I guess I better inform the Colonel of a potential relationship.”

“Oh, a commander's job is never done,” Christe said with a chuckle.

“So where would you like to eat? Here in the café or would you rather go to the flight shack?” Maxwell asked.

“How about we go back to my place. I have something I made the other day when you were out galavanting around the country.”

Well, it was decided off to Elizabeth's it would be.

While the pilots were learning about the XY-33 the additional maintenance troops that arrived from Edwards were getting their feet wet as well. Some had been sent over to the engine shop to assist with the production, others were getting hands-on with the test bird that Maxwell had taken up. They were going through it with a microscope, checking for any possible damage or fatigue.

It seems as though there was a new group of personnel every two to three days arriving, but all told, the last major batch was the Edwards group. There were still a few people due to be on station soon, one which was the love of Rogers, Michelle Grace.

As the day was coming to an end Maxwell released all of the pilots for the day except for two.

"Rivera and Davis, I need you two to stay behind there is a matter we need to discuss."

After all of the rest had left the room Maxwell began to explain.

"We are going to have four squadrons, right now I'm appointing each one of you as a squad leader. Tomorrow we will be breaking up into three teams. I will take a team to the engine cell, Rivera you will take a team to the XY-33 hanger, and Davis you will be with the last team in the Starboard hanger bay on the UE1. Rivera, you and your team will cover the XY from top to bottom. Listen to the maintenance troops, they have been working with it for several years and they are all top-notch. Davis, I want you to spend the day covering every square inch of the hanger bay. Make a plan for the most direct route from our living quarters and the other key areas of the ship. Do an alert drill and see how long it takes to get there, but stay out of the port side hangar bay. There are some major issues they are trying to work out and we don't want to impede their progress. After the lunch break, we will switch up the teams. My crew will go to the hanger bay and then Davis have your guys report to Rivera, so that would leave yours, Rivera coming to me at the engine shop."

Now let's go take a look at this beautiful masterpiece they have built for us to play with. That way you two will have a better understanding of it tomorrow when we cover the engine. Rivera, you will be on your own to check out the hanger bay, but I would get with Davis and see what they came up with for the route planning."

Both of them nodded their head in agreement at the same time they responded. "Yes, sir."

So the three of them set off to go get changed and meet back up outside of entry point one.

Chapter 11

March 18th, 2008, Nellis AFB, 1300 hours

Family members had been arriving for the past couple of days. There was a lot of confusion and concern among several of the families. They had arrived as they were directed to do but there was no sign of their spouses.

Colonel Starkes and Maxwell were greeting them at the base theater, directing them inside.

"You may sit in an empty seat." Maxwell was advising the people as they entered.

It was several minutes past the designated hour. Starkes took up a position in front of the theater to address the group.

"Good afternoon, I would like to welcome you to Nellis AFB. I'm sure there are many questions and some concerns. Let me reassure you that you will be joining your family members very soon. It will just be a little later today."

As each person was now looking forward they all stopped talking, giving their full attention to Starkes.

"My name is Colonel Starkes. I'm the commanding officer to which you're family members have been assigned. I'm sure there might have been some talk before they left but here it is in a nutshell. They are assigned to a top-secret project, one for the first time where the family will be allowed to go on. You will be leaving shortly to meet up with them. After your arrival, you will be going through a short orientation class to become familiar with the program and the nature of your involvement."

Maxwell walked over to the colonel and whispered something to him.

"It seems that our transportation is here. So if you will proceed out the rear door behind me two buses are waiting. As you leave Maxwell will be checking your name to the manifest. We don't want to leave anyone behind."

One by one each person made their way through the back door and boarded a bus. All in all, there were 114 children from a newborn to young teenagers, along with 47 adults. The bus ride out to Groom Lake was going to take about three hours. They would be arriving at about 1630 hours.

Now that they were underway Starkes and Maxwell headed back over to the transit office to get their flight gear and take the short hop back in the F-15. This would allow them to make final preparations for the orientation back at the base with their waiting family members.

As the buses were making their way down a very lonely stretch of the road, Groom Lake Road. They had gone through a security checkpoint. It had been several miles ago and yet they're still was no sign of a base, nothing resembling a base, actually there was nothing much at all to look at.

You could tell that the passengers were getting impatient, some were beginning to question where they were going.

It was at this time that an automated voice was heard over the income system on the bus. "*I would like to direct your attention to the right side windows you are now looking out to the area that is known as Groom Pass, you are now entering one of, if not the most restricted area in the United States, you are but a handful of civilians to ever see this location, on behalf of the President of the United States of America, welcome to Area 51.*"

Area 51 is what movies were made about, this was known as the UFO capital to the world, home of the Roswell crash and alien folklore.

One lady could be overheard making a commit. "We just seem to be out for a Sunday stroll through one of the biggest coverups that the world has ever talked about. This just doesn't make any kind of sense."

"I know what you mean, most of what my husband does he never talks about and here we are in the middle of it all now. Something does not add up. We are not being told something, and I have a feeling I'm not going to like it."

After hours of virtually no signs of life, upon the horizon, a little town was beginning to take shape. Wait that's not just a town, it's a military base, this place is real after all.

Both of the buses pulled up to the base theater. It was there that they were greeted by Maxwell and Starkes. Maxwell went into the first bus and Starkes into the second one.

"Welcome to your new home for the time being. Yes, you are at Groom Lake, or as some of you might know it by, Area 51. I have some good news for you all, inside you will find your family members. Just like you, they have been patiently waiting for your arrival. We have a short social event when you enter you will find a buffet line if you are hungry along with a beverage station and of course restroom facilities. You will have about thirty minutes to eat and visit before a short briefing will take place." Maxwell informed the occupants on the first bus.

Starkes was giving an almost identical speech to the individuals on the second bus.

One by one each person left the bus and entered the theater. Some went straight to the food, some were looking for their life partner and some made their way to the restroom. It was not long before everyone was reacquainted with each other.

Time moved fast and Starkes was once more upfront on the stage. Tapping into the mic to get the attention of the room.

“If you will all please find a seat we will continue with a short brief and what to expect over the next few days.”

It was not long before most of the room was seated and quiet. Of course, you could hear a child crying or asking questions to their parents. But it was mostly quiet, just like being at a church service.

“I know it’s been a long day but we have a few housekeeping items to address and then you will be dismissed. I know we have all ages in the room and some might not understand and that's ok. I’m here to tell all of you new arrivals that we are at a major milestone for the human population of this planet. We will be departing very soon on a grand adventure. One that will take you farther than your wildest dreams, one which most of you only thought was possible in a Hollywood movie. Turn down the lights please.”

Just on cue, the house lights dimmed and on the screen was projected an image of the UE1.

“On the screen, you are looking at Universal Explorer One, UE1 for short. Yes, this is real if you don’t believe me you can ask the person you are sitting next to. They will be able to reassure you that it is, in fact, real.

In fact, it is located right here. We will be leaving in a few months on this ship."

Starkes presses the remote on the podium and the screen begins to show a short movie clip of the ship and some of the major areas, like childcare, educational area, recreational facilities, the medical area, along with the hydroponics section.

"As you can see for lack of better words it is a small self-contained city. Each one of our primary members was selected for this program. Each one of you has been fully vetted and has the needed skills we need. We understand that the children are too young at this time, but they will receive the finest education in existence. With the ability to complete a college degree program of their choice. Now some of you are asking yourself where are we going and how long is it going to take? Well, the answer is about twenty years one way, and the location is a star called Sigma Draconis known as Alsafi we are hoping to set up a colony on a planet in its solar system. Your spouses have already committed for you to go on this grand adventure, but it is voluntary. If you choose not to go you will be required to remain here at Groom Lake until the launch. After that time details will be released to the population of the planet. You then will be allowed to leave this base. A small bit of insight that might help you make up your mind. Yes, there are alien life forms out in the universe, we are not alone and a huge amount of our technology for this vessel has come from recovered alien ships. You all have been asked to be a part of something larger than life, larger than anything mankind has ever known. We will give you some time to talk amongst yourselves. If you feel this is not for you then you can see Maxwell or me and we will make the necessary arrangements to have you removed from the program."

The room suddenly was becoming very loud as wives were speaking to their husbands and children were asking questions to their parents. Looking around you could see all kinds of emotions and facial expressions. Some showed excitement, some looked as if they had concerns, others were just recounting their childhood fantasies of being an astronaut or playing Buck Rogers. Well, it just got real.

A few individuals came up to Maxwell with a few more going over to Starkes. One lady was asking some questions. Not wanting out of the program, just looking to get some more answers.

Starkes went back to the podium, tapping on the mic to get the attention of the room.

"I know you all have more questions if there is no one looking to exit the program I will gladly answer some more for you now." Once more it became very quiet in the theater.

"We are planning a departure for July 4th, which is just a few short months away. This project has been underway for years. We will have communication portals available for you to speak with your families back here on Earth. We don't know how long it will work but we have taken steps to ensure the possibilities for complete communication all of the way there. Some of you will have the option to stay behind on a planet if we can set up a colony. It will once more be voluntary. The best course of action is to have you all speak together tonight in your living quarters. We will need all of you newcomers to meet back here tomorrow at 1000 hours. At that time you will be given a more in-depth orientation on the UE1, along with filling work assignments."

With the final message from Starkes, the room started to become empty. Maxwell and Starkes proceeded to leave as well.

"So how are things shaping up with our newly arrived aviators?" Asked Starkes.

"It's progressing very well, it would be great to get some real flying in though."

"Well, we should be getting a few delivered each week. As soon as the engine cores and fuel cells are installed you can have your guys do a fire-up test and taxi if all checks out we are going to have a mass launch and they can dock with the spaceport. I would select your top candidates as they will be staying on the station for the next few months."

Chapter 12

March 20th, 2008, Groom Lake, 2215 hours

Feverishly working his way through the checklist, Rogers was calling out commands just like a few weeks back.

"SMART run a full diagnostic, with priority on the past issues," Rogers ordered.

Smart responded with a basic answer. "In progress."

After what seemed like forever Smart responded with the words everyone was waiting to hear. “All systems check complete no detected abnormalities. All systems are in an acceptable range.”

“Rogers, did I just hear what I think I heard?”

“Yes sir, I do believe we have a clean bill of health.”

“Ok let's get a full report from each section chief.”

“Yes, sir.”

One by one Rogers was calling into each section. The response was the same, each chief confirmed what SMART already had. UE1 was ready for the next step, partial disassembly, and transport to Elmendorf AFB, Alaska. It was a short window to get all of the sections that would be going up in the Russian rocket launch to the ISS. Or as it was now known in this circle as Earth Spaceport, another section had to be transported to the Kennedy assembly building in Florida for a planned launch in May.

The complete assembly had to take place here but not fully knowing the effects of the gravity on Earth it was always planned to reduce the overall mass and weight of UE1 to get it off of the planet. Work was going to be non-stop and there was no room for error.

“Maxwell get a trip planned for Elmendort. We will leave Saturday sometime before noon.”

“Yes, sir.”

“Maxwell you have the bridge get her shut down and get everyone off. We will meet with section chiefs at 0900 in the main training room.”

“Yes, sir. Rogers begins the shutdown procedures.”

"Yes, sir."

"Con open a channel with all stations," Maxwell ordered.

"Your channel is open sir."

"Excellent job everyone. We are proceeding with the shutdown procedures, section chiefs will meet at 0900 in the training room. Once your station is powered down you are dismissed, bridge out."

"Rogers advise me as to when we are complete with the shutdown."

As Maxwell was looking about the bridge area, watching everyone work he began to wonder. *What an incredible wonder of technology has been assembled and here I'm right smack in the middle of it all.*

"Sir the shutdown is complete." Informed Rogers.

"Very good then, let's go get some rest and we will see you all at 0900 as well."

One by one each member was leaving their duty station, Maxwell was the last one out.

It did not take long for 0900 to come around. It looks as thou all of the section chiefs were seated in the room by the time Starkes arrived. Starkes and Maxwell entered at the same time.

"Let's try to keep this short. Most of you have been working for several years on this project. The time has come, we need to remove the housing pods and aircraft landing bays. Getting them ready for transport is our number one priority. The XY-33s will be flown and docked to the ISS. Each one is capable of carrying a limited amount of cargo. We need to identify and get it removed. Only the absolute bare essential items to maintain functionality remain with the ship. Along with securing all loose items we need to make sure all essential items are in the staging area to be sent for deployment on the Shuttle. Along with the pod it is carrying. The communication probes need to be tested, including the ones that have been sent ahead. Lastly, we need to have the main weapons magazine store loaded. Each of you has your standard assignments along with your secondary. The time to execute to 110 percent of your ability is now. We are nearly a fully functional spacecraft. Mankind's future depends on our success."

As was customary Colonel Starke's words of wisdom hit home with the crew. Outside of Rogers and Maxwell, most of the team had been assigned to the project for several years. It was almost unbelievable, time for partial disassembly.

"How are we looking for our departure tomorrow, Maxwell?"

"The flight plan has been filed. We have a couple of in-flight air refuelings to make but we should make the trip in about six and a half hours."

"And departure time?"

"1100 hours sir."

"What time do we need to show?"

"I'll be there at 1000 hours sir. As long you are suited up and ready to go 1045 will be fine sir."

"Very good then. Everyone you all are dismissed. Rogers, I need for you to hang back for a second."

Rogers stayed behind as the team left the room.

"Yes, sir."

"There's not much pressing that you should be needed for over the next couple of weeks. So how about you take a few days and go see your girl."

"Sir."

"Yes Rogers, I know we kind of pulled you out of your life and turned it upside down a bit. Take off and go see her, be back on the 26th."

"Yes, sir."

Rogers was off like a bolt of lighting. He was going to need to go pack a bag, try to get a flight out of Las Vegas, and get a rental car set up in Boston. MIT here I come.

Michelle was going to be blown away. With the way it has been for the last few months, Rogers had barely gotten in a few emails. If he was not working then he was sleeping, with working taking up almost all of his time.

Sitting at his desk reviewing the flights out of McCarran International Airport Rogers was feverously looking for the first flight heading to the East.

He found some doable options on Southwest Airlines, there were four departures, 1:40 and 1:55 were the first flights out that he could make.

Booking the one at 1:55 as it would get him there thirty minutes sooner, and every extra minute was going to be well worth it.

If you have ever seen a tornado blow then you could imagine the site inside of Roger's lodging facilities. Stuff was being tossed here and items were thrown there but before long he had two bags packed and was out the door.

As soon as Rogers cleared the main base area he was stomped on the accelerator. Seventy miles per hour, increasing to eighty and backing off when he hit eighty-five. Dust flying behind his car.

Normally this drive would take about three hours going the speed limit, which was posted at 55. Rogers was hoping to do it in under two and a half. That would give him just enough time to get to the Airport and get checked in for his flight.

With a light workday, and the Colonel off to wherever he may be Maxwell decided it was time to go for a visit with Elizabeth before he was off for who knows how many days.

Upon entering the training office, he found her sitting in her office working on the computer.

"Well, good morning stranger."

Elizabeth looked up, with a slight chuckle in her voice. "What brings you by."

"You know, just out for a walk."

"Some walk you got there that you ended up here."

Now it was Maxwell who was smiling and laughing.

"Thought I better get some one-on-one time with you. You know we are heading out again tomorrow, not too sure when we will be back."

"You and the Colonel?"

"That would be a correct assessment."

"If I did not know better I would think you two are seeing each other instead of you and me."

"It does seem that way sometimes, so how about you take the rest of the day off. I hear there is a great movie playing at the theater."

"Are you asking me to skip out on work?"

"Yes mam, I would exactly be asking such a thing."

"Well then, I hope it includes lunch?"

"Of course, it does my dear."

Elizabeth and Maxwell were off to eat and spend some much-needed time alone.

Starkes was in his office, speaking with Chavez.

"So how is the production proceeding Sargent?"

"It's moving along now, with the extra personnel that has arrived on station."

"Great, so what is the count?"

"There are three ready for transport and a crew from the 312th out of Travis is due in tonight. We will have them loaded and they should be ready for departure by 0900 local time."

"Ok, hold the departure till about 2100. That way it's dark and I will arrange to have some fighters from Nellis escort our mother ship back."

"Yes sir, I will see to it that the time is changed for a 2100 takeoff."

Starkes called down to Jeniffer.

"Get me the commander on the horn from the 57th ATG (Adversary Tactics Group), Nellis."

"At once sir." Was her reply.

After a few minutes went by.

"Sir I have Colonel Handy on the line."

Starkes reached over for the phone, line two was on hold.

"Colonel Starkes here, we are going to be tapping your unit pretty heavy the next few weeks. We are going to require your assistance in handling the matter on your end. I need for you to have a briefing with the selected pilots, they will be flying the same type of mission repeatedly over the next twelve plus weeks. It would be most beneficial if it were the same aircrews, the fewer people involved the better. They also need to hold a TS/SCI (Top Secret/Sensitive Compartmented Information) clearance, the project is codenamed, Caretaker. They will depart Nellis and proceed to Palmdale Plant 42. At that time they will escort the package from Palmdale to Tonopah Test Range. The first departure will take place tomorrow at 2100 from Palmdale. I know it's a bit late with notice and it might be an inconvenience but I hope you understand. SecDef's office will get the official orders sent down to you later today. Just giving you a courtesy heads up."

"Well thank you for your consideration, but if you don't mind sir what authority is this coming from?"

"It falls under Executive order 13458 like I said you will get formal orders from the SecDef, just thought I would be nice and give you a heads up. Trust me when I say you will not want to have any screw up's."

Starkes was not quite sure how Colonel Handy was receiving the information but was not going to waste much more time on the matter. As he knew, whatever he requested would be made official.

"Unless you have any more questions I have other pressing matters to take care of Colonel." So stated Starkes.

"No, I'm good."

So the two hung up the phone.

Handy called his vice commander to come to his office.

It was only a few short minutes.

"Yes, sir."

"We need to figure out who we want to send on an extended assignment, about 12 weeks long. We need to get some aircrews in place to carry out an exercise of escorting something from Palmdale to the Tonopah Test Range, and before you ask the orders are on the way."

"Yes, sir."

Starkes called back to Jennifer.

"Yes, sir?"

"Get Robert's office on the phone for me please."

It was not long before Starkes was relaying his plan and needs to Mr. Gates. Between the SecDef and the President, he was commanding a very large influence in the military chain of command.

Heading out of his office Starkes stopped at Jennifer's desk.

"If I'm needed I will be in my quarters for the rest of the day. The next few days I will be in Elmendorf if there is a pressing matter that needs my immediate attention."

"Yes sir, have a safe trip."

March 21st, 2008, Las Vegas, 1323 hours

Pulling into the parking garage, Rogers had no time to spare. Making his way as fast as he could to the terminal. Once inside Rogers went to the automated machine to retrieve his ticket, then over to the security line. Luckily, the line was short and seemed to be moving fast. Once through the security checkpoint, he was looking for his gate when over the intercom he heard. *"Last call for Southwest flight 133 paging passenger Michael Rogers, please report to gate B11.*

Rogers was now running through the terminal building, there it was B11. Out of breath at the counter.

"I'm Michael Rogers."

"Yes, this way sir. You almost didn't make it."

Leading him over to the agent by the ramp door. Which she was just about to close. Scanning his ticket and down the jetway, he went. Entering the airplane, he was greeted by one of the flight crew members.

"Welcome aboard, we are a full flight. I believe the last seat is near the back."

"Thank you."

Rogers made his way down the aisle, locating the last open seat. Of course, it was a middle seat but that was ok, for he now was going to be on his way to see Michelle.

Another crew member helped him with his bags as there was not much room in the overhead compartments by his seat.

There was a young lady in the aisle seat and an older gentleman by the window seat on his row. She moved out of the way so he could get into his seat.

"Thank you."

"Your welcome, I'm Marry. How is your day going?"

"Great, now that I made this flight. I'm Michael by the way."

"So where are you off to today?"

"Well, I finally get to go see my fiance. She is in Boston. It's been a few months and this is going to be a last-minute surprise."

"Oh, that is so sweet."

"What about you, where are you going?"

"Just going back home. I was just up here for a few days. I had a job interview at one of the casinos."

"So did you get the job?"

"Yes I did, I start in a few weeks. So I'm heading home to pack and get moved down."

"I just moved to this area not too long ago too."

"So what do you do?"

Rogers had to pause and think for a second.

"I'm a computer engineer, assigned to Nellis AFB."

"Wow, you seem so young."

"Well, you know how it is with the new generation. We are mostly computer and gamer geeks."

You could hear the engines starting and the air in the plane was switching over to the onboard systems. The flight attendants were making their final cabin checks before taking their seats.

The plane started to move back, being pushed from the ground tow cart. You could hear some odd noises throughout the cabin, sounds like the hydraulics moving the flight control surfaces. It was not long before the plane was moving on the taxiway, not much longer and they would be airborne.

March 22nd, 2008, Over the Washington Coast, 1353 hours

"Colonel we are coming up to our first in-flight refueling."

"So how exactly does this work?"

"It's kinda like pulling up to a gas station in your car, except we are pulling up to a flying gas station. This one is a KC-135 tanker to be exact. They will extend a boom that will go into our air refueling receptacle, located on the left side above the engine intake."

"So what if something goes wrong?"

"Well, then we pull that D-ring handle between our legs, and before you have time to kiss your butt goodbye we are ejecting out on our nice rocket-powered chair."

"No really what do we do?"

"I was serious. Colonel if something was to become unrecoverable we would punch out. I hope that the chute strapped to our back works. But don't think too much about it, remember flying is safer than driving a car."

By the time Starkes was able to run any serious scenarios through his head Maxwell was already making contact with the KC-135.

"169 to 51 moving into pre-contact position."

"Copy 169 I see you."

After a few seconds.

"169 move into contact position."

Maxwell slowly advanced forward and slightly up moving within a few feet of the tanker. The boom operator extends the lower part of the boom aligning it to the receptacle on Maxwell's jet.

"Contact 169."

"We have a lock if you could top us off."

"About 6000, sir?"

"Roger that, 6000 should be good."

It only took a few minutes to get the required amount of fuel onboard.

"Ok, we are full."

"Roger that sir you are clear to break contact. Maxwell backs away and starts to drop down and turn to the right, away from the tanker.

"Thanks for the top off."

"Good day, sir."

They were now about 100 miles away from the mainland heading over the ocean. One more meet-up to get a few more thousand pounds of fuel and a short hop they would be on the ground at Elmendorf AFB, Anchorage Alaska.

"Hope you brought some warm clothes Colonel. The report coming in, it is 32 degrees and it's going to be dropping to about 28 later this evening and into the early morning."

"I have a few items on hand. I'll be just fine."

As the hour was moving by it was not long before they had cycled through the second refueling. This would be the last time they would see the KC-135 crew as Maxwell would be increasing his airspeed. Now with enough fuel to complete the journey.

It was only about an hour more in the air and they would be back on the ground, one more leg complete.

March 22nd, 2008, MIT Campus, 0900 hours

It only took Rogers a few minutes to get reacquainted with the campus. Driving over to Michelle's dormitory.

Parking the rental car, it was almost impossible to hold back his emotions but Rogers was able to keep it all in check.

Knock, knock, knock is what Michelle heard coming from the area around the door. *Now, who would be beating so early in the morning on a Saturday?*

Got up from the couch to open the door, and at the same time she called out. "Who is it?"

"I have a private message for Ms. Grace."

"That would be me."

With complete astonishment, screaming out when she saw Michael at the door.

"What the hell are you doing here?"

"Why to see you of course."

"I know, but how, I mean oh hell, it's you, it's really you."

Michelle was just stuttering on a bunch of jibberish, not knowing really what to ask or say.

"I was given a few days' leave and thought I would come to surprise you."

"Oh my God, you don't know how surprised I'm, come in come in."

At the same time, Michelle jumped into his waiting arms. Michael had a grin from ear to ear, and Michelle had tears of joy going down her cheek. As the two of them entered the room and the door closed behind them. It would be some time before anyone else saw these two again.

Chapter 13

March 24th, 2008, Elmendorf AFB, 0930 hours

Starkes and Maxwell were to meet with the base commander at 1000 hours, General Tinsley.

Meeting in the lobby of the VOQ/VAQ both gentlemen decide to get a quick bite to eat from the continental breakfast provided for the guests.

General Tinsley was in his office, going over his daily schedule. Calling down to his assistant. “Sue, I see there is a meeting for ten on my schedule. Have that canceled I think I’m going to go to the golf range and play a few holes today.”

“Sir that might not be such a promising idea. It seems this request came down late yesterday from the Sec Def’s office.”

“SecDef, who is this meeting with?”

“A Colonel Starkes and Captain Maxwell. They flew in on Saturday according to the tower log, from Nellis. One of the transit aircraft that arrived late in the afternoon.”

“I don’t suppose there is any indication as to what or how long this meeting is to be?”

“No sir, just to clear the morning and to post a time of 1000 hours sir.”

“Very well then, I guess I’ll be sticking around for a while. Advise me as soon as they arrive.”

“Yes, Sir.”

As the time was approaching 0945 Starkes and Maxwell were having the last few bites of their meal. Outside there was a base taxi waiting for them.

“Where to today?”

“Base commander’s office.”

“Yes, sir.” Replied the driver as the car started to move.

It was not a very long drive to the commander's office. Starkes and Maxwell made their way into the reception area.

"So did you want me to call and have the M. Ps on standby to throw us out?"

"Funny Maxwell, very funny."

"I thought it was."

"Good morning, welcome to the 3rd WG. How may I be of assistance to you today?"

"Yes please, if you could direct us to the base commander's office."

"Proceed down the hall take the first left and you will see the reception lobby for General Tinsley. You should find his executive assistant Sue at the desk."

"Thank you," Starkes replied as they started to turn and proceed down the hall.

It only took about a minute to reach the General's office. Upon seeing Sue at the desk Starkes greeted her. "Good morning, Starkes and Maxwell we should have a 1000 appointment with General Tinsley."

"Yes sir, he has been expecting you. One moment please."

Picking up the phone and ringing the General's office.

"Yes, Sue?"

"Your appointment is her sir."

"Great send them in."

"You may enter now. Third door on the right."

"Thank you." Said Starkes once more.

Once entering Starkes spoke first. “May we close the door, sir? What we need to discuss is a private matter.”

“By all means.”

As soon as the door had closed even before Starkes, and Maxwell was able to get near the chair or couch.

“So, what is so damn important that my schedule had to get cleared? Besides keeping me from my golf game.” General Tinsley spoke with a bit of sarcasm and irritation in his voice.

“Well, sir not to burst your bubble. You may be living far up North. But you are not so far up that you, like the rest of us, have people to answer to. And in this case, it shall be me for today.”

“Excuse me, Colonel, are you trying to pull some attitude crap with me? I’ll be more than happy to make a phone call and have your caboose put right back on whatever plane you flew up here in.”

“You go right ahead and make that call. Because I’m pretty sure my call will be better than yours and your butt will be driving me around to look at whatever I wish. So do it, make a call, better yet save both of us some time and just call our Commander and Chief.”

“Are you saying to call the President? Now you’re trying to be a wise guy. I don’t have time for this crap.”

“Oh, wait a minute you probably don’t have his direct number.”

Pulling out his sat phone Starkes hit the speed dial key. It was only a few seconds before the phone began to ring.

“I’ll put it on speaker for you.

"Why is that you, Jerome?"

"Yes sir, George."

"Well, how are we looking?"

"It's looking, great sir. Maxwell and I are up in Elmendorf. Fixing to cover the necessary steps in operation Top Cover."

"That's some great news. I know that Ushakov will feel somewhat better. So, what can I do for you today?"

"Well General Tinsley just wanted to say hi and I wanted you to know that he is more than excited to be of assistance in this National Security matter. He even said he would drive us around himself."

"Great to hear General. It's always good to have such nice team players. Now you just try to grasp what my boys have in store for you. It's going to be a little touchy, but it's going to be well worth all the trouble. Now if there's nothing else you need, I have some other matters to attend to."

"I'm good George. I'll see you in a few days."

Looking over to the General, Starkes asked. "Are there any other concerns you would like to address before we continue?"

"No Colonel, I have a clear understanding."

"Good, then here's how this is going to play out. We will be sending some transports up here in April and May. The first two will be meeting with a Soviet transport one going to an Airshow in Canada, one when it returns. On these trips, it will require the need to refuel. So, it will be landing here under escort. We need for it to be parked in the most remote and least visible area possible. Our transport will be arriving in the middle of the night under armed fighter escort and be parked next to it. The flight line lighting will need to be off and only our M.P.'s will be in the area as cargo is offloaded from one aircraft and placed onto the other. This will once more occur in May, during the Icelandic Air Policing. An AN-124 cargo plane will declare an emergency and be permitted to land. Once more the transfer will take place in the middle of the night. The next day the so-called problem will be fixed and it will take off with armed escort till it is out of U.S. airspace and meets up with Soviet Mig's. Any questions so far?"

"There is no way that all of this is legal."

"General, this is so far above your pay grade and clearance you just have no idea what is taking place. The only reason we are here and you are included at all is your proximity to Russia. We need to seamlessly transfer the cargo from one aircraft to another with as few people involved in the process as possible. All you need to know is when it will arrive and when it is leaving your base. Also as you heard, the full support of the President and the SecDef are involved in this project."

"I'm telling you now that I'm not going to be some patsy in this game of chess."

"General, you will not be made into a patsy. But there will come a day you find out how important you were in helping this project along."

"So where is the originating point of this project?"

"Why Groom Lake of course. Now if you would be so kind as to show us the most remote and darkest place on the flight line we would like to see the area."

Walking past the commander's reception area.

"Sue, we are going to be out for a while."

"Yes sir, your schedule is clear till 1330."

"Thank you, Sue."

It was not long before they were in the general's staff car and driving on the access road for the flight line. Heading to an area off of Airdrop Avenue and adjacent to where runway 16 starts.

"This is probably going to be the most secluded area other than the Alert hanger, but there are people there 24 hours a day and they could be up in the middle of the night. Here you have three aircraft pads and it.s pretty much in the middle of the crossing of the runways. Also, there is no lighting on the tarmac here."

As the car came to a stop Starkes and Maxwell got out. Looking about to the left, to the right, there was not much around and it does seem to be away from the operations buildings.

The General got out to join them. "Well, what do you think?"

"It seems to be secluded. What about this building behind us?"

"That's just a ground support unit they are usually gone by1700, well before it gets dark.

You do realize that in May it's not completely dark for very long mainly from about 0200 to 0400 is the darkest time."

Starkes looked to Maxwell. “Damn, we didn’t consider that. Not that it would change much. We don’t have many options to complete the transfer. They're just going to have to work with a short window of time to get the task completed.”

“What if we call for a quiet hour?”

“I had thought of that, but having one designated at that time would just make more people curious and more likely to come out.”

The General now injected a comment. “You do realize there is a 2200 hours curfew on base already. The only aircraft authorized to take off and land is alert birds on official missions.”

“That might work to our advantage then, the Russian transport could already be in place. Ours could fly in right at 2200 hours. No one would be any wiser than our M.P.’s would take over the security while the cargo is transferred over.”

You could tell by looking at Starkes that he was becoming more relaxed, or better yet maybe a bit more relieved that the last few details were working themselves out. After years of secrecy, it was coming down to a few clandestine moves. Soon the final pieces of the puzzle would be falling into place.

“What about crew accommodations? We would like for them to be housed in the same building, away from other transient personnel. With the first transfers taking place over a week's time the crews will stay here on the post. On the third transfer in May they only need overnight lodging.”

"We have a newly renovated dorm that currently is not in use. How many people are you talking about?"

"Well, let's see, four ship fighter escort, so that's eight, a C-5 crew that would be 5 more, then we would have about twenty more in M.P.'s. Oh about thirty-three personnel plus the Russian crew. So say about forty should cover it."

"Are there any more requirements you are going to need?"

"No that should just about cover it. But let's have a look at the housing quarters you are talking about."

As the three men are getting back into the General's car, you could hear the alert horn going off. Looking over to the alert hanger the doors were opening on two of them. Less than a minute went by before two F-22 Raptors were on the taxiway heading to the runway. With a continuous roar and flame extending from the rear of the jets, they were airborne, with the afterburners still glowing until they were out of sight. The two jets would be hitting Mach 2 plus. Yes, they were in a hurry to intercept a bear. That is what it's known as. A *Bear Run* when a Russian bomber or any other aircraft is near our border we send up fighters. Just so they know that we know that they are there.

"So does that happen a lot," Maxwell asked?

"Well as a matter of fact. Yes, it is almost a daily occurrence. So the thought of transferring cargo onto their transport is just not making sense to me. Not when we send up interceptors all of the time."

“Well General, maybe, just maybe we are not that big of an enemy with each other, as one would be led to believe.”

“How do you mean?”

“Let's just keep it at what one is accustomed to is what one likes to accept as fact. And for you and your mission, you are to protect our Northern Border from any adversary. For you that mainly is Russia, and for me, it is something considerably different.”

With the three of them now in the car and on the way to the dorm building. Starkes was just looking to see that it would have all of the required necessities that his troops would be needing. A secure entrance, showers, bedding, and some nearby eating establishments.

After a short drive, they were outside of the building that Tinsley spoke of.

“Where are the mess facilities?”

“There are two in the walking distance one to the North and the other a few blocks over to the East going towards the BX / Commissary area. Along with a Burger King.”

“Very well, it seems that all will be in order. Speaking of food, how about we get something to eat?”

“Are you up for a round of golf?” Asked General Tinsley.

“A round just might be what is needed.” Replied Starkes.

“Good, then we can eat at the club. I think you will find the portions are substantial along with being quite tasty.

On the way over to the club, Starkes informed Maxwell. “File a flight plan for a Wednesday departure. Destination will be Andrews.”

“Yes sir, how many nights stay should I put for the VOQ?”

“We are not going to require any. We will be staying at the White House.”

Once arriving at the golf course and checking in for a tee time they went to eat.

“So Colonel, would you care to enlighten me some?”

“As to what, General?”

“How is it you have the President on speed dial? And you are on a first-name basis also? I can only think of a few Air Force officers that might have that info and you sir would not have been on that list. So what is it that you do?”

“You are so right, there are not many people that have a direct line to the President. With your position and time in the service, you should know when I tell you it is highly classified. It’s just the nature of our work assignment.”

“Hogwash Colonel, don’t take me for a fool. I know you have to be involved in some type of covert government project. One as to which we don’t wish to make public, like most of the early days of our newest experimental aircraft. What I don’t get is how the Russians are playing into all of this.”

“Well sir, it will all become clearer as the days go by. Now there is no need to get too involved with something you have no control over.”

“All I’m saying is I need some official orders to carry out this insanity. I’m not putting my neck out on the line if this goes south. I've got too many years in the service to risk my pension.”

“General, SecDef will have you covered. There will be a dispatch of coming events very soon.”

Once the meal was consumed it was almost their tee time. Waiting in the pro shop was short-lived as they were called for a few minutes later.

The golf game went pretty uneventful, the General held a two-stroke advantage going into the fifteenth hole when he received a call.

“Sorry to bother you but your 1330 appointment is waiting for you, sir.”

“Oh well, we are almost done. I should be there about 1430.”

“Yes, sir I will advise them.”

Finishing up the last four holes the General held onto his lead and won the match by two strokes over Starkes, Maxwell was a bit further down by five.

Well if there is nothing else I’m needed for, where can I drop you off?”

“The VOQ will do just fine.” Informed Starkes.”

As Starkes and Maxwell were entering the building Maxwell made a slight joke.

“See I knew I should have placed the M.P.’s on call. You were going to get us kicked out again.”

"I don't seem to be able to help myself. It's just my nature. You can file the flight plan for tomorrow. Just leave word with the desk on our showtime. I've got a few things I need to see to. Besides you can use a day off as well."

Chapter 14

March 26th, 2008, Above Montana, 1017 hours

How many more of these air refuelings are we needing to make?"

"Two more after this one. Today would have been a good day to use the XY-33. It would have made short work of this cross-country trip."

"Yes it would have, but it would be a bit hard to keep it from being noticed. Although we seem to be making good time."

"Well, we are cruising just under Mach 2. Unlike our trip to Alaska where we used the KC-135 to ferry us we are meeting up with different tankers along the flight path. This allows us to maintain a higher speed for longer periods. We only slow down to make contact with the tankers and take on fuel."

"So why didn't we do that for the flight to Alaska?

"Well it's not SOP, plus we are going through 4 time zones on top of a normal flight time of 9.5 hours. This allows us to cut the flight time down by four-plus hours."

As the next few hours went by and the last tanker stop was complete, it would not be long before they arrived at Andrews AFB. What Maxwell was not aware of yet, the President had Marine One standing by to bring them to the White House.

As they were now on the final approach Maxwell had received proper clearance and landing instructions from the tower.

As the jet turned off of the runway, they began to follow a transient truck to their designated parking spot.

"Look there Colonel, it's the Presidental helicopters."

"Why yes, I do believe you are right."

"I wonder why they parked us so close?"

As Starkes and Maxwell got down from the cockpit, Maxwell met up with one of the ground troops. Giving him a quick de-brief and filling out the maintenance logbook. Starkes had gone over to the travel pod to retrieve their bags.

"It was a good flight. I'm going to write up one of the fuel cells, it seems that the gauge was sticking a bit or hesitated as we were taking on fuel."

"We will look into it, sir." Responded one of the transient airmen.

Maxwell joined the Colonel in retrieving his bag. At this time they were approached by another pilot, a Marine officer.

"Good afternoon, welcome to Andrews. I take it you are Colonel Starkes and Captain Maxwell?"

Starkes answered yes we are."

"Great if you two are ready, follow me and we will be on our way." Spoken as he turned and proceeded to go to one of the waiting helicopters.

"What the hell?"

"Oh it's nothing Maxwell, it flies just like the Jets do just not as fast."

"I get that part, I was just not expecting to be going in it."

Starkes began to laugh, with a broad smile on his face and patting Maxwell on the back.

"I get it, we are here to meet with the President, well at least you are but this is pretty cool. Elizabeth is never going to believe this."

"We will be taking off very soon. So if you will just fasten your seat belt we will be on our way."

It would only be a short trip from Andrews AFB. Flying out over Maryland and D.C., on their way to the White House. Not very many people get to see Washington from this perspective.

Upon entering the estate of the White House, they were met by some secret service members. Getting escorted into one of America's most famous buildings.

"Welcome back Colonel. Captain, I don't believe we have had the pleasure of meeting yet?"

"No sir this is my first trip here." Informed Maxwell.

"Well, I hope you enjoy your stay. My name is Troyer, you're going to need this while you are here."

Handing each one a badge to wear while on the premises.

"The president has been expecting you. I'll take you up to the Oval office."

Following Troyer through the maze of halls, it was not long before they had arrived. There was a presidential secretary outside next to the oval office. She called the President. Upon lowering the phone she informed. "You may enter now."

Next to the door was stationed another secret service member. The door was closed behind them when they entered.

"Jerome it's so good to see you. So I presume this is Captain Maxwell?"

Starkes replied. "Yes Mr. President, this is the outstanding pilot that I have been telling you about."

The president reached out to shack Maxwell's hand.

"It's an honor to meet you, sir."

"No, the honor is all mine. If it were not for this small detail of being president I would be going with you on this marvelous journey. One that you will soon be leaving on. So have you told him why he is here?"

"No sir, I felt that it should be you."

"Tell me what?" Maxwell asked.

The president went over to his desk to retrieve a small box.

"Captin Maxwell, if you would please stand."

Doing as the president asked Maxwell stood up as well as Starkes.

"Captain it is my great honor to present to you the insignia and the rank of Major, effective yesterday the 25th of March."

"Thank you, sir, I don't quite know what to say."

"Nothing needs to be said, your work and the great mission that lies before you speaks for itself. Colonel Starkes, would you please pin on the insignia, but first, let me get the photographer."

Picking up the phone and pressed a button to ring his receptionist. The president called to have the photographer enter. It only took about thirty seconds before Pete the official White House photographer entered.

Starkes was pinning on Maxwell's new rank while the President shook Maxwell's hand as Pete took several photos of the event.

"Very well, Colonel Starkes it is with great pleasure and honor to present to you the rank of Brigadier General. As the President shook his hand Pete once more took the picture. Then the president proceeded to change out Starkes insignia and replaced it with a star.

"Well sir, you pulled a fast one on me. I was not expecting that at all."

"That might be so, but it is long overdue. It took a bit of time to get the Senate Armed Services Committee approval."

"Pete put a rush on the photos, have them placed into a nice frame."

"Yes sir, they should be ready later today. Where would you like them sent?"

"Just send them to my secretary she will get them to where they need to be, thank you."

"Ok, now that we have the formalities over with, how about getting me up to speed on our progress."

As the three of them sat down to cover the events that took place in Alaska and the latest milestones on the UE1 you could see that the president was very pleased.

"I must say Jerome when I was first briefed on this project I never thought we would see this day come. Now here we are only a few short months away."

"I would have to say the same George, as you know most of my career has been dedicated to this project."

"You know the nation, hell the whole world is going to be following your adventure, most of them wishing they were up there with you."

"Well sir, we will do our best not to let anyone down. With the team we have assembled, we have some of the smartest people on this planet working damn near around the clock. You can feel the energy and excitement every time you enter a room."

As the three of them kept talking time seemed to stand still, oblivious to life outside of the room. As the conversation went well into the night the little party broke up at about 2230 hours. The president went to his private sleeping quarters while Maxwell and Starkes were led to a room of their own.

As Maxwell was laying on his bed, his mind began replaying the day's events. A long day of flying and several hours visiting with the president, and a promotion to boot.

Before long it was morning. Not wanting to be late Maxwell was up early, dressed and ready for the day. Not sure where to go Maxwell went to Starkes room. Knocking on the door.

"It's open."

"Good morning sir. I was not too sure where to go."

"We are meeting the president and some of the senior staff for breakfast. Then we have a closed-door committee meeting with the Armed Services Committee. After the briefing, we will conclude the day back here. We will depart Friday morning, back to Groom Lake."

March 27th, 2008, Russell Senate Office Bldg., 1045 hours

"Yes, mister chairman that would be correct."

"So, you are trying to tell me that we have spent over 500 billion on developing this program?"

"Yes, sir."

"Now then what are we going to get out of this pipe dream?"

"Well for one we hope the survival of the people of this planet. You are aware more than most sir, we are no longer on the top of the food chain."

"Are you trying to be condescending with me?"

"No sir, just relaying the facts."

"Then please explain your reasoning."

"We have been chasing alien encounters for the better part of six decades. You have seen the reports, some of the incidents have been very hostile, to say the least. The consensus is once they learn how to survive the entry into our atmosphere there will be no way to stop them. Currently, we have no way of knowing their intentions."

"When will the UE1 be complete?"

"We have begun the final phase. Partial disassembly is taking place as we speak. Sections are being relocated for transport to the Space Port. We are prepping for a July 4th departure from the planet. At that time, it will dock with the spaceport. Providing all goes as planned it should take about ten days to complete reassembly."

"So, then I presume your departure then will be the middle of the month?"

"Yes sir, we are scheduled for an early release from the spaceport at 0800 Zulu. At that time there will be a final worldwide disclosure. Along with a video feed of the main engine ignition and departure of the vessel."

"Are there any other questions from the committee members?" Asked the Chairperson.

After a short pause, the chairmen looking to his left and then to his right was satisfied.

"Well general, you may go with the blessing of this committee and God's speed to you all."

Now with the closed-door session complete, Starkes was free to head back to the White House. It would be a short drive only about ten minutes or so.

Chapter 15

April 21st, 2008, Groom Lake, 1730 hours

The past two weeks, just like the preceding months have been remarkably busy.

Outside of the UE1 hanger, a C-5 cargo plane was parked. The C-5 had arrived earlier in the day. Having been filled with some of the UE1 section pods.

The aircrew was making their final checks before departure. The window of arrival was short, they would need to be in Anchorage by 2200 hours.

Once there the support crews would need to offload the precious cargo and get it onto the waiting Soviet Antonov AN-124. The Soviet transport had also arrived earlier in the day.

It was a special goodwill measure to refuel the aircraft, as it was to be displayed at the Canada Air Show. Well, that is what the public was led to believe.

Maxwell was meeting with ten of his newly acquired pilots. Having attended many briefings before, this was Maxwell's first as commander.

"We will have a standard 2 by 2 take off with the second wave of 2 by 2 departing at a two-second delay. The last two will execute in a single ship formation. Over the Tonapah range, you will all increase to ten percent military power and make for low orbit. At that time Ghost will make contact with the spaceport and receive docking instructions. Any questions?"

There were no questions asked.

"Ok then, crew showtime is 1845 hours with departure at 1915 Zulu, dismissed."

Once more the night flight of the fireflies would be out in full force. Ten XY-33s were being transferred to the spaceport. Along with the C-5 to Alaska, Maxwell was to fly cover.

Maxwell made haste to an awaiting F-16. He was already in his flight gear, along with his go-bag having been placed into the travel pod earlier.

"169, Tower, over."

"Tower, 169 you are clear for engine start, over."

As Maxwell was powering up his jet so were the other three fighters that would be going on the trip. Just as Maxwell and the others were completing their flight control checks the C-5 started to Taxi. Once the C-5 cleared the area they would also be on the move to the EOR.

The time was 1800 hours when the C-5 began to roll down the runway, followed by the four F-16s. It would be about 6 hours in the air with two scheduled fuel transfers to take place.

The flight was mostly uneventful. With the last cycle of the fuel, the fighters were now all set to make it to Elmendorf AFB.

The C-5 would be the first aircraft to make an approach and land. Maxwell would be the last one down.

With the C-5 parked near the AN-124, the crews would be able to make short work of the transfer. Just like Starkes and Maxwell planned a few weeks back. The F-16s were close by but a little farther down the ramp.

Base Transient crews picked up all of the aircrews. Half of the MP's that had arrived with the transport stayed with the Aircraft while the others went to get checked in to their temporary living quarters.

About forty-five minutes later the MPs returned so the others could go get checked in.

Most of the flight line area was dark and pretty quiet. The time was approaching 0145 hours. The support crews returned to start transferring the cargo from one aircraft to the other. They had a short window of time to get the assignment complete. This time of the year the sun starts to rise at about 0430 hours. Once it became daylight it would be too easy for their work to be spotted by someone.

Getting the modules transferred took less time than they anticipated, having completed the work by 0335 hours.

The team was making their way back to their temporary living quarters. Ron was the first to speak up about wanting something to eat. Most of the crew was in agreement. Luckily for them, there was a 24-hour chow hall on base. So first they would stop at the chow hall, then off for some rest.

As the sun was coming up, the base would soon be coming to life. Today promised to have a bit more excitement in store for the residents of Elmendorf. There would be a huge turnout for people to see the Russian aircraft. Little did they know what was inside.

Maxwell and the crew of F-16 pilots would be on escort duty as the cargo plane took off. Not so much for what was inside but more for the local people to feel secure.

Crew show time was set for 0830. Takeoff would be at 0900 for the first aircraft. Once they were in a position to provide top cover the transport would take off. Followed closely by the other three F-16s.

There was a called quiet hour from 0800 to 0930. A quiet hour means most activities on the base were placed on hold. The only activity taking place on the flight line would be in direct support of the departure of the Russian aircraft.

All of the designated aircraft were in position at the EOR.

"173 you are clear to expedite departure, over."

"Tower, 173 copy, over."

173 increased the engine power just slightly, as the jet began to move onto the runway. Having the breaks applied 173 went into full afterburner. With the release of the breaks 173 shot forward with a thunderous roar. A short way down the runway Stricker pulled back on the flight controls and the jet lifted off of the runway surface. Shooting straight up into the sky. It was not long before 173 disappeared into the clear blue sky.

"Tower, AN-124 you are clear for departure, over."

"AN-124, Tower copy, thank you for your hospitality, over."

The AN-124 transport moved onto the runway. Without stopping they just kept rolling, increasing power and speed. They traveled much further down the runway before they began to lift off. As soon as they had started to move two F-16's took up position on the runway.

The AN-124 had cleared the runway surface and was proceeding to increase in altitude. The next two in position were given the clearance to depart.

Maxwell was in the last aircraft. He now moved into position on the runway. Once more you could see the flames coming from out the back of the engine as he went into full afterburner. Soon he was gone from eyesight.

Over the skies of Alaska, all four of the F-16s were protecting the AN-124. Soon they would be meeting up with some Russian fighters.

About ten minutes into the flight first contact was being made.

"11 o'clock, 2 pairs of bogies, over." 173 called out.

"Copy, 173. Go make contact it should be our handoff. The codeword is *Swift,* the response is *Recovery*, over." Maxwell instructed.

173 was still on the top cover, about 5,000 feet above the rest of the aircraft. 173 increased speed and turned to intercept the approaching aircraft.

He was about 50 miles out when he began to try contacting them on the designated radiofrequency.

"Unkown aircraft this is the US Air Force, you are approaching US-controlled airspace. You need to make a Swift Departure"

"US aircraft we are trying to make a Recovery as we speak."

"Copy Recovery, Welcome to the Bering Strait."

"And to you my comrade."

"The package will be here in about five minutes. Standby while I inform them of your arrival."

The four Mig's began a holding pattern, along with 173.

"Cosmo, Stricker, I Have a visual on Recovery, over."

"Copy, Stricker, We are five minutes out, over."

"Copy, Cosmo, five minutes, over."

It was not very long before the transport plane was being handed over to the Mig's for the escort back to the Cosmodrome. Soon the secret cargo would be on its way to the Space Port. Once there it would be waiting for the arrival of the mother ship, UE1.

Now that the escort duty was complete for the day, Maxwell and the others turned back for their temporary home at Elmendorf AFB.

Over the next few days, Maxwell and his team would be flying some training missions. The C-5 would leave near the end of the week. Once more it would go get more modules, for soon there would be another AN-124 aircraft here to receive the cargo.

Chapter 16

May 23, 2008, Earth Space Port, 2135 hours

Ghost was in the command center watching the monitors. Ten more XY-33s were due to arrive any minute. He and the other nine pilots had now been on station

for just over one month.

The first few days were spent learning how to work in a spacesuit when not in the pressurized areas.

Once their ships were empty of the cargo, they brought with them they had a lot of spare time on their hands. Each crew member was assigned an additional job duty to perform. For the most part, they worked about four hours, and the other part of the day they would fly training missions. With no commute to work, this still left a lot of downtime.

Looking out a window Ghost saw the first approach of the new arrivals.

“Flight lead, to ESP, over,” Jackson called out. “ESP, lead you are clear for docking station 4, reduce to impulse power, over.”

“Roger ESP dock 4, over.”

One by one all ten XY’s docked with the spaceport. Once they were on board their cargo would also be moved to the central hanger bay.

Ghost met up with them at the docking station. He would be showing them to their living quarters for the next few weeks.

“Welcome to the ESP. I hope your flight was uneventful?” Ghost asked.

“I must say it was a unique experience. I never thought I would be flying in outer space.” Francis responded.

“It is quite different, my first time was an incredible experience, to say the least.”

“So, what’s the first order of business?”

“Well, I will show you to your living quarters, then we can take a short tour of the station.”

"Great, I don't know about the rest of the team, but I could go for something to eat."

"We can get something in the chow hall after we are done."

May 24, 2008, MIT Boston Massachusetts, 1000 hours

Graduation day was finally here for Michelle. It was to be a very hectic and packed eventful day. As soon as the ceremony was over Michelle was to be checked out of her dorm and she was to proceed to the airport for a flight to Las Vegas. Not understanding the need to go to Vegas, Michelle was confused. Having chosen to work with the FBI Cyber division, she thought it would have been in the Washington area.

With the time nearing early afternoon Michelle was hurrying about, most of her room was already packed. Just needed to get the last of her clothes and travel bags.

Waiting outside for the taxi she called for, Michelle tried to call Michael. Every time she dialed his number the phone would ring one time and immediately go to voicemail. The odd thing was it just felt different, not quite able to put a finger on it she just blew it off.

Once more looking out into the street, this time she saw a taxi approaching. Walking up to the car the driver called out. "Are you Michelle Grace?"

"Yes, I'm."

"Great, I had a little trouble finding this building."

"It's ok, but I do need to get to the Airport, I have a flight at three."

"No problem, you will be there in twenty minutes."

As they were driving off Michelle was looking around, reminiscing, and thinking of the past four years here. Looking back, it all seemed to go by in a flash.

Before she knew it, they had arrived at the airport.

"Which airline are you on?"

"Southwest."

"Very well, Southwest gate it is."

Pulling up next to the curb, the driver got out to help remove her suitcases.

Her ticket was already printed, so she was able to go straight to the security line. Once in the line, it was only a few minutes before she reached the clerk who was checking the tickets and ID. Next to the X-ray scanners, placing her bags on the conveyer belt and small personal items into a small tray. She was directed to walk through. Collecting her items, she was off to find her departure gate.

Sitting down at the departure gate, Michelle once more tried to call Michael. Just like the last few times, one ring and straight to voicemail.

People were already in line to board the plane. Her boarding pass was numbered A12. She did not see a point to stand in a line when you already have a position number.

As the time approached 2:25 pm, one of the Southwest employees made an announcement. “Flight 2672 to Houston, then on to Las Vegas now boarding group A.”

Thought to herself that was my cue to get in line.

It was going to be a long day, arriving in Vegas at 8:30 pm, and that includes two-time zone changes.

Placing her carry-on in the overhead bin Michelle then settled into the seat she picked. Opening the book she brought with her, she began to read *The Hunger Games.*

May 24, 2008, Earth Space Port, 1300 hours

The Russian Space Transport (RST) was due to arrive in the next few minutes. One team was stationed at Docking Station 2, another was in the main cargo bay. The command center was a beehive of personnel.

This was another major milestone in the UE1 project. What began as a dream from one country has turned into a worldwide endeavor, upon a global scale of cooperation never seen amongst our population.

“ESP to RST we have a visual on you, you are cleared to proceed to dock station 2. Just follow the flashing blue marker lights, over.”

“RST copy, proceeding to Dock Station 2, over.”

"Once you make contact you will have a soft dock, marker lights will turn red. When you have the green marker lights you will have a hard seal and clear to enter, over."

"RST copy, soft dock on red and hard dock on green and clear to enter, over."

The ESP has its chain of command. While Ghost has temporary command of the flight crews.

"Commander, would you like any assistance from me or my team?" Ghost asked.

"Not at this time, but when we begin the removal of the UE1 pods we could use your help in the cargo bay. Also, we received a communication from Groom Lake for you."

"We are happy to help any way we can. Where can I view the communication commander?"

"The LT can pull it up for you on his screen."

"Great, thank you, sir."

Ghost made his way over to where the LT was stationed.

"The commander advised me that you would be able to pull up the communication dispatch for me."

"Yes, sir. Just type in your access code and it will display on the screen." Replied the LT.

LT typed a few commands into the terminal, then moved to the side. Ghost put in his access code. Immediately the screen changed and displayed the following message.

Groom Lake command advises the following. If all goes as planned, you should be receiving a set of pods from the UE1 today. On the next delivery of XY-33s, there will be a group of technicians. We understand that living space might be getting a little cramped, but they will be staying in the modules that should have arrived today. The techs will be working on the reassembly of the components that have arrived with the XY deliveries and the other transports. Attached to this message is an XY mission plan. As you now should be familiar with the XY, we are staging an extended flight. One to encompass speed and duration. Follow the specs and select three crew members including yourself for a total of four. The destination is the moon. We need to record the exact time to reach the moon's orbit. Make sure you run the mission clock; we need to have accurate data to complete engine valuations. You have two days to get the XY's ready. The only way to retrieve you if something goes wrong is with the RST while it is still at the station. The RST crew has already been notified, as soon as their ship is offloaded they will refuel for an extended flight if they need to recover you. Good luck and God's speed.

Ghost spoke out. “Well, this is getting real now.”

“Everything ok?” Asked LT.

“I guess we are going to the moon.”

“Seriously?”

“Yes, it seems in two days.”

May 25, 2008, Las Vegas, 0835 hours

Michelle had stayed at a local hotel for the night. She was to call for a cab that would take her to Nellis AFB. Along with the set of instructions that had arrived by courier a few days ago was a civilian ID. Once she got to the visitor's center, she was to inform them that she needed to go to building 828.

Waiting in the lobby, the taxi Michelle called for had arrived. Walking outside with her bags, she informed the driver of her destination.

The ride out to Nellis only took about 25 minutes. The taxi proceeded to drop her off at the visitor center.

Once inside she was greeted by an airman.

"Good morning, welcome to Nellis. How may I help you today?"

"Yes, I'm scheduled to go to building 828."

"Yes mam, I need to see your orders and ID, please."

"Orders? I don't think I have any."

"Ok, just let me see your ID."

Michelle handed over the ID she had received. The airmen checked it against a roster he had. Finding her name on the list. There was a notation that she was a civilian, so she would not have any orders.

"Yes mam, you are expected this morning. If you like you can have a seat, there will be a shuttle shortly. You can take it to building 828, just let the driver know when you board where you are going."

"Thank you."

Michelle went over to some empty seats near the door. Not knowing how long the wait would be, she once more pulled out her book to read.

About 10 minutes had gone by when the shuttle driver entered the room and called out.

“Base shuttle will depart in five minutes.”

Picking up her bags and leaving the building, she saw the shuttle. Walking up to the door, the driver greeted her.

“Good morning. Where are you going today?”

“Building 828 please.”

The driver wrote down her building number on his clipboard. Two more people also entered the shuttle. One was going to 903 and the other was going to the base BX.

After a few more minutes two more people got on the shuttle. One was also going to 828 and the other was going to 1600.

Michelle heard the person mention building 828. So, when they came down the aisle she spoke up.

“Excuse me. Are you going to building 828? Are you here with the cyber division too?”

“Cyber division, what’s that?”

“I took a job with the FBI, and I was sent here.”

“No, I’m here to meet with my husband. He is on TDY.” (Temporary Duty Assignment)

Michelle was looking puzzled. TDY, not sure what that is. Why am I here, she thought?

"What's a TDY?" She asked.

"TDY is a temporary duty assignment."

"Oh, how long has he been here?"

"About two weeks. It was different than the other times he has gone on a TDY. He came home one-day acting kind of strange. He packed a bag and was gone the next day. Usually, he would have some advanced notice."

"Really? I'm Michelle by the way."

"Nice to meet you I'm Judy. So, I gather you are going to 828 also?"

"Yes, that was the building listed on the letter that was sent to me."

"Letter? You mean your orders?"

"No, it's a letter, remember I'm with the FBI."

"Oh yeah, now that seems a bit strange."

"Why is that?"

"Well, this is an Air Force base, FBI training is in Virginia. Well, I'm pretty sure it is, anyway."

"That's what I was thinking."

The shuttle came to a stop in front of a building. The driver called out. "Building 828, 903 is across the street."

Both Michelle and Judy got off the shuttle, along with a guy who went across the street.

Walking into the building they were greeted by a clerk at the reception counter.

"Good morning. How may I help you today?" Judy spoke up first

"Yes, I'm here to meet my husband, he is on TDY."

"And your name please?"

"Judy Michaels, my husband is Adam Michaels."

Looked over a roster, finding her name. She would be going to some off-base location.

"Yes mam, go down this hall room 104, there will be a briefing shortly for you."

"Thank you."

"And you are?" Looking at Michelle inquired the clerk.

"Yes, I'm Michelle Grace with the FBI, I was to report to this building."

Once more looking at the roster. There was her name as well.

"Yes mam, you are to go down the hall to 104 also."

Thank you."

Michelle proceeded down the hall while thinking. *This seems kind of strange, I'm going to the same room as Judy.*

Once entering there were a few people in the room including her new acquaintance, Judy.

"Judy, I see we meet once more," Michelle called out.

"Well, I suppose so. Now, this seems odd." Replied Judy.

"I was thinking the same as I was on my way here."

As the two of them sat down to talk, it was clear they were becoming friends.

About five minutes had passed before Starkes and Maxwell entered the room. They both went up to the front of the room behind the desk.

"Good morning, everyone. This is Major Maxwell, and I'm General Starkes. I have a short briefing before we get on our way."

Looking at the people in the room you could see a lot of confusion on their faces.

"Most of you are joining your spouses on their deployment, and some of you are here on assignment."

Michelle was pondering the last statement. *I'm not so sure I'm here on an assignment, and surely not joining my spouse.*

"I'm going to be very straightforward with you. We are very near to a monumental event. One that the world has never seen. One which you have been invited to be a part of. Each one of you has a role in this endeavor. First, I must advise you that if you choose not to partake you will be confined to a secure location for about 6 weeks. Second, most of you have a spouse that has already committed to this project. The few of you who have been recruited to join, we need you on this project. As it is too far along to try to replace you. If someone will dim the lights, please.

Starkes proceeded to turn on the projector. Displayed on the screen was a picture of the UE1.

"Ladies and gentlemen, I present to you UE1, this is to be the world's first deep-space transport ship. It is fully self-contained. We have been constructing it for several years. Our planned departure is in 6 weeks. July the 4th. So now you will begin to understand. If you choose not to go, we cannot have this information divulged before we depart the plant. That is the reason for the confinement. Yes, alien life forms are real and yes, they have been here on our planet for some time.

Once more looking into the eyes of the people in the room. It was hard to tell if it was disbelief or amassment. But one thing was for sure. The curiosity level just went up a few notches.

"Before I go on, is there anyone who feels that they will not be going? There will be no repercussions for you or your spouse, but they have already committed."

Michelle raised her hand.

"Yes, mam." Said the general pointing to Michelle.

"I'm just a bit confused. So, I'm not hired by the FBI?"

There were a few chuckles heard around the room when Michelle made that statement.

"No Michelle, but Michael is a part of this team. And believe me, is very much waiting to see you."

Thinking, he knows my name and Michael.

No one else raised a hand.

"Very well. For those of you that are married, we hope you choose to expand your family. Those that are single we hope you find love amongst the crew. As there are several thousand people on the team, with a sizable portion who are not married. We are traveling several years away in search of a livable planet. If there are no pressing questions, please hold them for your orientation class, which you will be attending tomorrow."

With a short pause just to make sure the questions did not start rolling in.

Judy raised her hand.

"Yes, mam."

"Where are we going now?"

"Groom Lake."

"Where is that?"

"Maybe you know it by Area 51."

"Oh.

"Ok, Maxwell go check to see if our transportation is ready.

Maxwell left the room to check on the shuttle bus.

Almost immediately the small talk began.

Maxwell returned a few minutes later. Looking to the General, giving a thumbs up.

"Ok, if you will follow Maxwell, we are ready to depart.

As they left the building, there was a base shuttle ready to take them to the base transient center.

It only took a few minutes to get out to the flight line by the base transient center. There was a C-130 standing by. This would be their ride out to Groom Lake.

May 26, 2008, Groom Lake, 0900 hours

All the new arrivals were taken to the base theater for orientation. As most of them were civilians, the orientation would be much different than the other groups Elizabeth had been doing.

"Good morning. I'm Elizabeth Ely. As you are now aware we will be leaving in a few short weeks. Today I'm going to cover the major sections of UE1. Also, we will be assigning your job assignments for those that do not already have one."

As each person was seated and listing to the instructions being given, one could only imagine what it was going to be like.

Chapter 17

May 26, 2008, Earth Space Port, 1000 hours

Ghost was in launch position along with Boomer and Big Show.

"ESP, Ghost you are clear to depart."

"Copy ESP. Ok, Boomer take the lead, proceed to 1000 miles, and hold."

"Copy."

Boomer slowly exited the Space Port.

"Alright, Big Show take it out to 2000."

"Copy Ghost."

Big Show was taking up a position behind Boomer. In less than a flash, Boomer was out of visual from the station. Now Big Show disappeared just as quickly.

Ghost was outside of the station pointed in the direction of the moon.

"Ghost to ESP ready to proceed."

"ESP copy. We have a good track on you, whenever you are ready."

"Copy"

Ghost sat there a few minutes, looking over all the indicators and the HUD. (Heads up display)

"Ghost to Boomer. Are you ready?"

"Boomer ready."

"Ghost to Big Show. Are you ready?"

"Big Show ready."

"Ghost to ESP. Start the count."

"ESP to all stations ready on the clock. Five, Four, Three, Two, One, Ignition."

Ghost hit the throttles, full Military power. In less than a blink of an eye, he was gone. Boomer barely saw a flash as we went by. Next was Big Show. Big Show never saw Ghost go by.

"Ghost to ESP."

"Go ahead Ghost."

"How is the tracking?"

"We are having a tough time with the calculations, we lost you about thirty seconds in."

"Oh."

Ghost was breathing a little hard. Looking at the moon it was becoming larger and larger as the minutes went by.

There was no direct radio communication to Area 51 from the ESP, so they would have to send updates through computer messages.

Starkes was sitting in his office looking at the monitor on his desk.

Finally, a message came in.

ESP reports Mission to the Moon is underway. Ghost departed at 1006 Zulu, we lost him after 30 seconds.

Upon reading the message Starkes yelled out a few profanities.

Jennifer heard the loud shouts from Starkes office knowing it was out of character for him. She went to see if he was ok.

"Sir is everything ok?"

"Ah damn, it looks like we lost one of our birds."

"What do you mean sir? Nothing has left the base today?"

"No Jennifer, not here. From the ESP."

"Oh."

As she turned to leave. Starkes sat back down and began to type.

Is there any chance of recovering the individual and the craft? Who died on the ship?

Starkes sat back waiting for a response. It takes a few minutes for the signal to reach the satellites and for the ESP to be able to read it.

LT was on the communications terminal. “Oh damn, sir we have a slight problem.”

“What is it LT?”

“I don’t think Area 51 understood your message.”

“How do you mean?”

“Well, sir they think someone died.”

“What?”

“Yes sir. I think they feel that there was an accident. They are asking who died.”

“Oh hell. Send this response. Sorry for the confusion. Mission to the Moon is still underway. We lost the tracking on the ship about 30 seconds in. We have no reports of a mishap, repeat no reports of any mishaps. We have received voice confirmation from the XY’s all is still good. He was just too fast for our tracking systems.”

Starkes was pacing about the room. Too upset to sit still. Never has he lost anyone under his command. About five minutes had gone by when his computer made a beep sound, indicating he had a new message.

Reading the latest, he once more yelled out.

Jennifer once again came into his office.

“Sir?”

“They are trying to give me a heart attack.”

“How so?”

"The test is still underway; they just lost the ship on their tracking."

Now the two of them were just looking at each other when Starkes began to laugh, followed by Jennifer.

"Ghost to ESP."

All Ghost was getting back in his headset was silence.

"Ghost to Big Show."

"Go for Big Show."

"Can you reach the ESP?"

"Let me try. Ghost to ESP."

"ESP copy."

"Ghost is trying to reach you. Do you receive him?"

"Negative Big Show. He must be out of our communications range."

"Copy ESP."

"Big Show to Ghost."

"Go for Ghost."

"They are not receiving you. You must be out of their range."

"Copy. Advise them my speed is currently at 79,303 mph and still proceeding up slowly. Also, you better move out another 40,000 miles and have Boomer move out to your position."

"Copy."

"Big Show to ESP."

"Go ahead, Big Show."

Ghost wants Boomer to move to my position and I'm to move to 40,000. Also, he is reporting a speed of 79,303 mph and still increasing slowly."

"Copy Big Show. Check-in when you get there."

Big Show set a trajectory towards the moon and proceed to increase his throttles to move to 40,000 miles out from the ESP.

Boomer was the first to arrive at the new location and checked in with the ESP.

"Big Show to ESP."

"Sir it sounds like Big Show, but it is too broken to understand."

"Have Boomer confirm."

"Yes, sir. Boomer ESP."

"Go for Boomer."

"We are not completely receiving the transmission. Can you confirm if it was Big Show?"

"Affirmative, Big Show said he was holding at 40,000."

"Sir, Boomer confirmed it was Big Show he is at 40,000."

Starkes picked up the receiver and paged Jennifer.

"Yes, sir?"

"Can you bring some coffee to my office?"

"Yes sir, at once."

Once more Starkes computer made a beep sound indicating another incoming message. Looking at the monitor.

The reported speed is 79,303 and still going up. Had to reposition the XY's. Ghost is out of our communication range and so is Big Show. Big Show is 40,000 miles from the station. We can get broken messages from him but not clear. We tried to run some calculations but having lost him 30 seconds into the test all we have is inconclusive data.

Jennifer was entering the office with the coffee when Starkes jumped up from his chair and once more yelled out. While being startled she almost dropped the coffee she was carrying.

"Sir, you need to stop yelling like that you're giving me a heart attack."

"I'm sorry Jennifer, but this update is just, well it's just very good news."

"That's good sir. So do you think I can get some work done now?"

"Yes Jennifer, you should be able to."

The XY test was still going to take a few more hours to complete. Although the small amount of data and information being relayed was very encouraging. There would be a few days more back on earth to make a positive conclusion and study the data, but overall, it was looking incredibly positive.

Chapter 18

June 28, 2008, Groom Lake, 1034 hours

Groom Lake was completely closed. No outside personnel or transports were to enter the base before the departure of UE1.

For the past few days, all the individuals were going through complete records and medical checks. Identification cards were given to the non-military members along with shots and other essential needs.

It has been a few weeks since all the pods were delivered and transported to the ESP. As much of the re-assembly as possible had taken place. All that was needed now was the main ship. It was less than a week and the countdown was underway.

Displayed in the main hanger was a huge countdown clock. If one were to look at it. It would show 6-20:15:10. Hard to believe only six more days.

Michael was showing Michelle around the Bridge. You could not separate the two of them. Not that anyone was going to try.

"If you were to have told any of this to me, I would never have believed you, not in a million years," Michelle stated.

"I know how you feel, ever since I was first approached back in basic my life has been going crazy."

"I keep thinking I'm going to wake up and find myself back in school."

"Well, my dear, it most certainly is not a dream, but we may be living one."

Looking over the ship, you just got the feeling it was different. Not that some of it was already in space, something more like it was alive. It just had the feeling as if it were a living breathing object.

When you looked throughout the hanger, it too was different. For one there were a lot fewer people moving about. As all the construction and disassembly work had been completed most of the personnel had been released. This was their time to prepare their personal items that would be going with them. Each person had a total limit of 86 pounds that they would be allowed to take with them. It was just not a weight restriction. All your items had to fit into a metal container which was 2' x 4' x 18". Each person had a designated day and time to bring their personal box to the staging area.

Work tools, spare parts, and mission equipment items were already on board and their weight was accounted for. Once that was taken into consideration it was determined that all members would be allowed the 86 pounds, regardless of position and rank.

At the end of each day, the storage boxes were transferred into the UE1. The crew was small only about twenty people. Which meant each person had to move about 50 containers per day. Most of them would be placed into the hanger bays. Some would be outside of your living quarters for the families, while the four-man pods would have the container placed in the room.

Not all the living space had been assigned yet. That was one of the tasks to take place once they got to the ESP. One of the reasons was the fact that over half of the living quarters were already at the station.

Personnel like Starkes and the command team had been assigned their quarters. Some of the crew like the Engineering section and the Hydroponics also had there's.

The days seemed to go by very quickly, yes there are still 24 hours in a day. But here it felt like there were only 13.

June 30, 2008, Staging Area, 1354 hours

Michelle and Michael were in line to get their containers checked in. Once reaching the front Michelle was directed to place her box on the scale.

“Mam you are over by four pounds. You will need to discard some weight.”

“How do I do that? You know how hard it was to get to this point?”

“Trust me, I have heard that line a few times already.”

“Just move your container, I think I will have some spare weight in mine.” Stated Michael.

Michael placed his on the scale.

“Sir you are at 83 pounds, so between the two of you, you need to drop one pound.”

As Michael picked up his container they stepped out of line. As the next person was placing his box on the scale.

“Do you have any books or magazines?” Michael asked Michelle.

“Yes, I have a few. But I need them. I need everything in there.”

“You can toss the books, when we get to the station, I can download it into your I-pad.”

“What?”

“Yes, I will just download the books from the internet, and you will have a file on your I-pad that will have them available for you to read.”

“You can do that? Never mind, I’m sure you can.”

Michelle removed three books. As she got back in the front of the line and placed it on the scale.

"Mam you are now at 89 pounds."

Michael placed his container back on the scale, it still weighed 83 pounds.

"Ok, you two have a combined weight of 172, you are good to go."

There would be one more day of personal boxes being brought over to the staging area. Then the crew would begin the boarding.

It was going to take two days to process everyone. Those with quarters would be the first to enter along with mission essential personnel. Not that everyone wasn't essential, but some more than others.

July 3, 2008, Staging Area, 1454 hours

Personnel had been arriving most of the day. Each person was given a time slot on when to report. For the most part, it was mission-critical personnel first. Also, anyone whose living quarters were previously assigned on the UE1 would be boarding today.

Most of the A shift team for the Bridge were already on board when Rogers and Michelle arrived. Rogers had instructions to begin the power-up procedures at 1500 hours sharp.

First, they would go to their room and secure their container. This was the first time that they would be in their room.

"Wow, it's smaller than I had envisioned." Stated Michelle.

"Yes, I have to agree with you."

As the two of them looked it over, it was about a 12' x 12' space. Pretty open not much in the room. There was a desk in one corner and the other had a door. Looking behind the door, it was the restroom area. It was about half the size of the main room.

"Where are we supposed to sleep?" Asked Rogers.

Michelle went over to the backside of the room. On the wall was a control panel. Pressing one of the buttons. A lower door opened, and the bed lowered down from the wall.

"You didn't know?" She asked.

"No, I did not get to go through a full orientation class as you did."

Michelle started to laugh.

"What else do I need to know?"

"A lot more than we have time for right now."

Michelle and Rogers started to head to the Bridge, time was approaching 1500.

Back out in the staging area

As more personnel were arriving. As soon as there were twenty-five members ready, they were to be escorted to the entry point. All the entry doors were sealed. Only the one on the Northside was being used for personnel to enter.

Each person would go through the body scanner and ID check. Once in the locker room. Each person went to the counter as their ID was scanned, they would be given a bag and jumpsuit which would be colored to their responsibility. Men went down one side and women went to the other. As they left it was down to the shuttle car through the elevator. Everyone was instructed to proceed to the Auditorium. Once more each person would be checked in. Every movement, every detail, each minute was mapped out. There would be a briefing which would take place every thirty minutes.

Afterward, they then would go to board the UE1.

Onboard the UE1

Exiting the elevator, Rogers and Michelle proceeded the short walk to the entry point for the Bridge. Once at the doorway, Rogers swiped his ID card through the scanner. As they were entering you heard an automated voice over the intercom. *OIC Rogers on deck.* OIC stood for the officer in charge.

SMART was designed to recognize which officer was on the Bridge and authorized to execute commands. As a failsafe once SMART becomes fully online, there always had to be a commanding officer on the bridge. So as soon as Rogers enters the commands into the terminal it would be so.

Rogers went over to his station; Michelle was seated next to him. Both put on a headset. This would allow them to hear the response from the computer without the rest of the Bridge being disturbed.

Rogers looked around the room. “Well, I guess it’s showtime.”

Rogers began to type on the console. *Execute Presidential Order 10398678010318223300556.*

Pressing the enter button the first step of history was just written.

“Allegra advise Ground Control we are ready to initiate the first stage of power-up,” Rogers called out.

“Yes, sir.”

Allegra turned a few control knobs on her console, then pressed her talk switch.

“UE1 to Ground.”

“Go for Ground.”

“We are proceeding with stage one power-up.”

“Copy UE1.”

“Sir, Ground has been notified.”

“Thank you, Allegra.” Stated Rogers.

Once more pressing the intercom switch on his console, Rogers was now speaking to the whole ship.

“All stations we are initiating power up at this time please stand by.”

Letting go of the switch, Rogers was just speaking to SMART now through the headset.

“SMART turn on the internal auxiliary power.”

“Auxiliary power on,” SMART replied.

“Comm, EVO how do you read?”

“Comm is green and showing Aux power.” Replied Allegra.

“EVO is good also.”

Rogers pressed the switch on his console for the Engineering section.

“How do you read Engineering?”

“We are good sir.” Replied the Engineering Chief.

“Comm have ground kill the power feed.”

“Comm to Ground Control.”

“Go for Ground.”

“Kill the power.”

“EVO power up your systems.”

EVO flipped the main toggle followed by the four switches. One for each oxygen generator.

“EVO when you are at 100% inform Comm so we can disconnect the external air feed.”

It was not very long before EVO showed 100% and the external air was disconnected.

“SMART run a complete system check, exclude the removed section pods from the report.”

“Diagnostic test underway.”

After about twenty seconds SMART responded with the results.

"Check complete, no abnormalities identified."

Rogers looked at Michelle. "Well, that is a huge relief to hear."

"Yes, it is, the commander should be pleased with the result."

It would still be several more hours before all the designated personnel were aboard for the day. Tomorrow the boarding process would be repeated for the remaining personnel. As the T-1 day was ending, so was the blindness to the world.

Chapter 19

July 04, 2008, White House, 09:15 hours

The situation room was full of people. Sec Def was present along with the Secretary of State, Director of the CIA, and the FBI. Most of the cabinet members and a few of the other administration staff were also in the room.

As the President entered the room, the room became quiet with everyone standing.

"Please, be seated." Stated the President.

As the cabinet members took up a seat at the table the junior staffers were seated along two of the sidewalls of the room.

The president was looking toward the SecDef when he began speaking. "Mr. Gates, if you would read them in, please."

Looking at the personnel in the room you could see some puzzled looks. Most if not, all would already be up to date on the current events of the world. Knowing that nothing major had taken place overnight left them to ponder what was going on.

"Dim the lights please."

It was only a few seconds after the request when the lights were dimmed.

Mr. Gates pressed the controller for the projector. Displaying on the screen was a photo of the UE1.

"This is Universal Explorer 1, or UE1 for short. This spacecraft will be launching later tonight. There are approximately 4500 souls on board, the destination is Alsafi. While it has been in development for the past decade, out at Area 51. Along with a new space fighter craft, which will accompany the ship."

Gates changed the screen to show the XY33.

"This is the XY33, after a few test flights it has already been pre-deployed to the space station, which is a tad bit bigger than you might think or recall."

The screen now changed to show the ESP.

"This is the International Space Station, which is now called ESP, Earth Space Port. As a few of you are aware, there have been many Alien encounters that have taken place around the world over the last few decades that we know of. Well, a lot of this technology has come from these alien escapades, dating back to the infamous Rosewell incident. After the launch, tonight at 23:00 hours the president will immediately go live with a broadcast on all of the major channels. As we have no way to hide a ship this large during the departure process the president will inform the public at that time. Our partner nations which include Russia, China, and the United Kingdom will follow suit and inform their population at that time as well. The FAA will be clearing the skies beginning at 18:30 hours, there will be no departures or arrivals on any aircraft of any type. All military units in the CONUS will stand down at 18:00 hours. Only the alert units will be active. There is to be a coded alert transmission at 22:55 hours placing them on temporary hold. The directive will be given as a national security test. As you can gather from the size of UE1 we don't think we can just move out of the way very quickly, not to mention the amount of unimpeded airspace we are going to need. We don't need some Cessna flying into our 500 billion dollar spaceship."

Looking about the room the expressions on their faces were all over the board. The look of total disbelief. As they looked at the president, you knew it was true.

"Now it goes without being said, but I'm going to say it anyway. There is to be a total 100% blackout on this. We don't need a bunch of crazies trying to get close to the site. Or worse some fanatical group trying to shoot it down. As a preventive leak and security measure, you are all to remain in the security protocol until after the broadcast tonight. Okay, I think we covered all of the pertinent information for now. Now we all have some maters that we need to attend to, to prepare for the event, so let's get with it." Stated the President.

Talk about a bombshell, Mr. Gates and the President just dropped not one, but a handful.

July 03, 2008, Groom Lake, 11:00 hours

The remaining personnel were going through the boarding process. Most of the individuals today were not mission-critical. Of course, there were always exceptions, like Maxwell and Elizabeth. Starkes would be the last to enter.

Rogers was down in the Engineering section meeting with the Chief.

"Can you explain the power fluctuations last night?"

"We don't have an explainable reason right now."

"How long was the anomaly?"

"There were two recorded incidents. One at 22:03 which lasted for five seconds, and the other was at 03:30. This one was much longer, it went on for seventeen seconds."

"What specific details of the event do you have?"

“The 22:01 event power spiked to 101%. The 03:30 event was the opposite, power dropped to 98%.”

“Start looking through the power demands and subsystems. We need to identify something, and quickly. I’ll be in the Ready Room if you find something.”

As Rogers was making his way to the door he turned back to the Chief.

“Isn’t there reduced power settings at 22:00?”

“Yes, I believe there is.”

“I would look into that first.”

As Rogers continued to leave thinking to himself. *That might explain the first incident, but the other and more extended in duration.*

As Rogers was walking to the Ready Room, he took out the SMART communicator. “SMART run a diagnostic on the Auxiliary Power Converter.”

It did not take long before there were two beeps on his handheld. Picking it up and pressed the talk button. “What are the results?”

“Test complete, system fully passed.”

Well, of course, that would be too easy.

Making his way to the Ready Room, Rogers knew he would need to contact Starkes.

Picking up a phone receiver, depressing a button for Comm.

“Communications Allegra speaking.”

"Allegra this is Rogers I need to speak with Starkes. Can you try to reach him?"

"Yes sir, I'll try."

Disconnecting the call with Rogers, Allegra called to the ground station."

"UE1 to Ground."

There was a short pause.

"Go for Ground"

"We need to speak with the General."

"I don't see him anywhere."

"You have a landline; call him this is important."

"Copy."

Eddie called out to the room. "Does anyone have a number for the General?"

"Call the Tower, they can reach him." Someone said.

Eddie picked up the landline and dialed the tower extension.

"Tower, Airman Martinez."

"This is Eddie with UE1 Ground, we need to reach the General."

"Well, he's not up here."

"No but you have a number or a way to reach him, don't you? The UE1 needs to speak to him."

After what seemed like forever, Starkes was with Ground Control.

"Ground UE1 I got the General."

"Copy Ground."

Allegra called to the ready room. Rogers was still the only one in there. Picking up the receiver.

"Rogers."

"I have Starkes patching him through right now."

"Hello, can you hear me, sir," Rogers called out.

"Yes. What the hell's going on?"

"We had an issue last night with the power readings. One time it spiked to 101%, and several hours later there was a drop to 98%."

"What are the thoughts of the team?"

"Chief is looking at the possibility of reduced power settings for the first incident. But there is nothing solid for the power drop."

"Well damn, could this keep us grounded?"

"I don't know, I just don't know."

"Ok, keep working on it. I should be there at 16:00."

"Yes, sir."

Rogers was reading through lines of code, trying to find anything that might explain the cause.

After becoming frustrated and feeling hungry, Rogers went to the mess hall to get something to eat. Sitting at one of the tables Rogers was joined by one of the mess halls cooks.

"You, ok? You seem like you are disturbed by something."

"No, I'm just trying to work through an issue."

"Anything I can help with?"

"No, not really, unless you can explain why we had a loss of power in the middle of the night."

"Oh, you mean like when you turn on a light switch."

Rogers picked his head up, with a slight grin on his face. "Yeah, just like when you turn something on. Damn, you might be on to something, thanks."

"Glad to help."

Rogers took off, forgetting all about being hungry and his meal.

"Hey, you're not going to eat?"

Turning back and grabbing the sandwich that was on the tray.

Back in the Ready Room. Using the intercom for the ship. Ding, ding two tones went over the speakers throughout the ship. "All department heads report to the Ready Room. Repeat all department heads to the Ready Room.

People began arriving about five minutes later. Overall, it took about fifteen minutes for everyone to get there.

"Thank you all for getting here so quick. Last night we had a couple of power fluctuations. Chief what have you found?"

"The first issue appears to be the reduced power settings. It just took a few seconds for the system to compensate."

"Great I had a feeling you were going to say that. I'm pretty sure the second one was caused by the same effect. Just this time there had to be some large power draw on the system. What has me mystified is the time.

It took place at 03:30, as most of us, would have been asleep. Any thoughts?"

One hand went up. Rogers pointed at the person. "I'm sorry I don't know your name yet."

"I'm Eric. I work down in the Hydroponics section."

"You're not the department head?"

Just now realized his coveralls were a different color.

"No, sir. Richard is not here yet."

"Do you have anything to share?"

"Yes, sir. Each morning at 03:30 the solar growth lamps are programmed to come on."

"Every morning?"

"Yes at 03:30."

"Well, Chief?"

"It's possible. Last night was the first time we were disconnected from external power. So, the sudden demand on the Auxiliary Power Converters could have dropped some before the system compensated for the need."

"Alright, let's research it. Does anyone else have any thoughts or input?"

Everyone started to leave, going back to their duty stations. Chief, Michelle, and Rogers stayed behind.

"I think the best way to test it is to set the power conditions to 03:30 and have the lamps come on."

"Okay, so how are we going to do this?"

"My best thought is just to make an announcement and have everyone do like they did last night at that time."

"Let's plan to try this in thirty minutes. Does that give you enough time to get set up Chief?"

"I should be good with that."

"Okay, well let's move then."

Chief took off for the Engineering section while Rogers and Elizabeth went to the Bridge.

Once at his workstation, putting on his headset, Rogers called over to the Comm section. "Comm notify Ground to hold all personnel from entering the UE1. We need to run a controlled test."

"Yes, sir." Replied the Women on the communications panel.

It did not seem like much time had passed when the Chief called up.

"We are already in Engineering."

"Okay, thank you, Chief."

Rogers pressed the switch to communicate to the ship. "All stations, we are going to perform a critical power test. We need you to set the conditions for how it would have been at 03:30 last night. So, if your workstation was powered down last night, we need for you to do it at this time. Any device that uses the ship's power you need to turn it off currently if it was off last night. Please stay in place and do not move about the ship until we give the all-clear."

After waiting for about five minutes, Rogers called to the Chief.

"How is our power reading?"

"Something must still be on, something with a large power use?"

"I'll make another announcement."

Once more pressing the switch for the ship's intercom.

"All stations, you should be holding in place, do not turn on any power. The only power consumption taking place is what would have been on at 03:30 last night."

The comm panel made a tone and a light came on from the Chow Hall.

"Communications Jade speaking."

"The dishwashers and the ovens are running; we are unable to turn them off at this time."

"Hold please."

"Rogers, the Chow Hall has an issue with their equipment. They are not able to turn them off currently."

Rogers called down to Chief.

"The Chow Hall is running equipment."

"I think we are good. I have a stable reading and we can monitor for any fluctuations."

"Okay then, stand by."

Pressing the headset controls. "SMART turn on the solar growth lamps."

After a few seconds.

“Lamps on.” Was the reply from SMART.

Rogers changed the intercom back to the Chief.

“How’s it looking?”

“Good, I think we found our problem. I have the same amount of drop-in power and the duration was almost the same.”

“If you are confident, then so am I.”

Once more making an all-stations announcement, Rogers gave the all-clear to resume normal operations.

“Comm inform the Ground we are done, and they can resume the boarding of the personnel.”

As time began to slip away you could gather the emotions and energy of the crew. The anticipation level was reaching a new high, soon, very soon they would be skyward bound.

Rogers would soon be meeting with Starkes in the Ready Room, to brief him on the events from last night and today.

Chapter 20

July 03, 2008, UE1 Ready Room, 16:35 hours

Only Starkes, Rogers, and Chief were sitting at the table. Going over the events of the last twenty-four hours.

"Okay, so both of you are in agreement. The Solar Growth Lights were the root cause of the power disruption."

Both Chief and Rogers shook their head while both said yes.

“Is this going to be a continuing issue that we need to have concern for?”

“Well, I don’t think so. Last night the ship was only running on the Aux power. With the external feed cut off and our main engines not yet online, it was just a small hiccup. It’s not like it went into a critical condition. The power only drooped down to 98%, which was only for a few seconds.” Stated the Chief.

“Well, we need to be 100% sure.” Said Starkes.

“Once we get to the ESP tonight the main engines will be idle. So, we will have several days to test and monitor the systems before we depart.” Replied the Chief.

“Will the data be accurate? We are going to have crews working non-stop with the reassembly process.”

“Until we have all systems online and full power generating, it is possible to see slight fluctuations. But I don’t feel there is any cause for concern to worry about.”

“Very well then, we are still a go. Let’s get on with the depart prep and prepare to launch.”

Chief left to go back to the Engineering section. While Starkes and Rogers proceeded to go to the Bridge.

Once back at the Bridge Starkes took up his position in the captain’s chair. Rogers was at his terminal station. All workstations were covered, even the ones that were not yet powered up.

“Comm, check with Ground and get a status on the ship pull out.” Said Starkes.

“At once sir.” Replied Allegra.

“UE1 to Ground Control.”

“Go for Ground.”

“The commander wants to check the status of the ship’s pull-out.”

“Stand by for an update.”

A brief minute or two went by before the Ground Control called back in. One would imagine the few individuals still working in the building would be going a mile a minute.

“Ground to UE1.”

“UE1 go ahead.”

“Report to the commander all is on schedule. The rollout will take place as planned at 19:15 hours.”

“Copy, thank you.”

“Sir Ground reports they are on schedule for 19:15 rollout.”

Rogers was running some more diagnostics on the systems. The Navigation tech was just sitting at the terminal. Navigation would not get powered up until after the rollout. The optical scanners need to see the stars and the sun to coordinate the position. Looking about the Bridge it would seem that the Comm was doing most of the work.

“Sir I have a conflicting report.” Stated Rogers.

“What’s the issue?” Asked Starkes.

"Well, I just ran a headcount. It seems we have a discrepancy on the ship's roster."

"How so?"

"The roster count is 4824. Well according to SMART, we have 4825 onboard."

"What! You got to be kidding me with all the security procedures we have in place. How the hell are we going to find out who this extra damn person is?"

"Commander." Called out Allegra.

Yelling back to her Starkes did. "What!"

"Medical just called, we have our first baby born on the ship."

"Well damn. What else are we going to have before we even get off the ground?" Starkes mumbled out.

Starkes pressed the all-call button on his console. "All stations this is the commander, it appears that we have had our first birth, congratulations to the parents. We are approximately two hours out from our planned hanger departure. Take this time to double-check that all of your designated sections are secure. Commander out."

Two hours went by in a flash, it did not even seem like minutes had gone by.

"Commander, ground is ready to begin the rollout," Allegra informed.

"Good, inform them to proceed. Open the window shield."

While Allegra was communicating with the ground controllers to commence the rollout procedure, Rogers spoke into his headset. "SMART open front window shield."

As soon as Rogers gave the command the shield began to open. For the first time, the crew members on the bridge were able to see out of the ship. Looking out of the window all could see the display on the wall read 0:00:03:03.

"All stations we are three minutes out, you should now be secure in your designated station for movement." Stated the commander.

The hangar lights were turned off, the massive door began to open. This was the first time in over ten years, a few people were holding their breath that they still would work. Looking out from the bridge all you could see was the time, now showing 0:00:00:20. Green flashing arrows started to flash out on the ramp. There was a slight jerk motion as the tow tractors started to pull out the massive ship, it was not moving much faster than a snail's pace.

For the people that were on the outside, the most amazing thing was taking place. The birth of UE1 started to appear from the hanger. It just kept coming and coming, there seemed to be no end. It took over two minutes for the UE1 to clear the entryway. Once cleared the two tow tractors made a forty-five-degree turn to the left. It would take about twenty-five minutes to get the ship in position at the end of the runway. Once there they would disconnect and clear the area.

“Ok, Nav turn on your station. It’s going to take some time for the computer to track all the known stars and planets to place our location.” Rogers informed.

“Powering up now.”

“Advise when you have a star map on your screen.”

“Allegra let’s try out our communications, now that we are clear of obstructions. Get the NASA controllers online.” Commander Starkes called out.

UE1 was outfitted with a vast array of antennas and communications devices. UHF, VHF, GPS, Satellite, and Lasar Point.

Allegra established contact with Mission Control at the Kenedy Space Center in Houston.

It was only earlier in the day when Mission Control was read into the mission plan.

“Commander I have the MC on the line.”

Starkes already had his headset on. “This is commander Starkes of UE1. Who do I have the pleasure of speaking with?”

“This is Jerry Jason I will be the flight director. I must say I was completely surprised when I received the call today. No one here had a clue that there was anything of this magnitude in the pipe, much less ready for departure.”

“Well sir, then I am pleased, we have done our job well.”

“We have you set for a 20:00 hours local time launch. Is that still holding?”

"Affirmative, we are in position and completing power up at this time."

"Copy, we are standing by waiting for you to clear the ground."

"Understood, will be back with you shortly, UE1 out."

"Star map is up on my primary display."

"Commander we are green across the board, recommend we fire up the main engines." Replied Rogers.

Starkes turned his comm over to the all-stations setting. "All stations, this is the commander. We are fifteen minutes out, repeat we are fifteen minutes out, all stations should be secured at this time. We are about to depart on the biggest event in the history of humanity. A great president once said a day will live in infamy, well tonight will live in the heart and soul of everyone on the planet."

Starkes turned off the intercom.

"Lieutenant Rogers, commence full startup."

Rogers called engineering. "Chief fire up the core."

"Copy engaging the core." Replied Chief.

Less than a minute had passed, and a green power light was displayed on each workstation. Rogers was taking notice. "Propulsion initiate idle."

"Set to idle, firing up main engine one. One online, starting two."

Looking at the propulsion tech he was moving some knobs and lifting a toggle switch. "Two online."

The view from the hanger flight line area was incredible. Looking at UE1 the glow from the rear of the ship looked just like the reddish-orange glow of a sunset.

"Main Three online." Replied the Propulsion Tech.

Rogers looked to Starkes. "Sir, she's ready to go."

"Maxwell, prepare to engage flight controls." Calmly said, Starkes.

"Comm transmit for our final clearance," Maxwell asked Allegra.

Allegra had NASA and Groom Lake controllers on the line.

"UE1 ready for departure and requesting clearance."

Groom Lake responded with clear to move at your discretion.

"Houston ready and the skies are clear coast to coast."

"Sir we are clear."

Starkes once more turned on the shipwide address system. "All hands we are clear for immediate departure. Maxwell, four, three, two, one, engage."

Maxwell applied engine thrust and took hold of the steering wheel controls.

There was a sudden forward movement as the engines began to produce an incredible amount of energy and propel the ship down the runway.

Pulling back on the yoke, UE1 was clear of the surface of the runway. The overall speed did not appear as if the ship was moving at all, but looking from the ground what was very large was now becoming smaller and smaller to the naked eye. In a flash, they were gone from view.

"UE1, Houston, we have you on the scope and the clock is running. Safe travels my friends, your first destination may just be a few minutes away, but from here it is a lifetime, God Bless…

Email the author
elytm@yahoo.com

Did you find any of the easter eggs?

Follow on Facebook
https://www.facebook.com/TMEly.UE1

Enjoy the first book in the series...

UE1: The Secret Revealed

Coming soon

UE1: The Journey

&

The “ART” Series

A Special Ops team was sent to recover confidential material throughout the world.

Made in the USA
Middletown, DE
19 May 2022

65958539R00166